The Destroying Angel
The Rifle-Musket as the First Modern Infantry Weapon

BRETT GIBBONS

With foreword by

Lieutenant General Phil Jones CB CBE DL
The Royal Anglian Regiment

Copyright © 2019 Brett A. Gibbons

All rights reserved.

ISBN: 978-1-719-85727-7

The opinions expressed herein are those of the author, writing in a personal capacity about a subject of professional military historical interest, and do not represent the position of the United States Government.

DEDICATION

To Lt. Colonel Michael S. Schwamberger: Soldier, Leader, Mentor

Foreword

1	Introduction	1
2	Introduction of the Rifle-Musket	8
3	The Day Battle Changed	18
4	Inkerman and Sevastopol: A New Kind of Battle	33
5	Fire Tactics: A Paradigm Shift	45
6	The Rifle and the Trained Soldier	63
7	Support by Fire	81
8	The Rifle in the Indian Rebellion, 1857	95
9	Meanwhile, In France…	115
10	The Rifle-Musket in the Civil War	138
11	The Austro-Prussian War 1866	155
12	Conclusion	180

Poor lord! is't I
That chase thee from thy country and expose
Those tender limbs of thine to the event
Of the none-sparing war? and is it I
That drive thee from the sportive court, where thou
Wast shot at with fair eyes, to be the mark
Of smoky muskets? O you leaden messengers,
That ride upon the violent speed of fire,
Fly with false aim; move the still-peering air,
That sings with piercing
  Shakespeare, *All's Well that Ends Well*

FOREWORD

I joined the British Infantry in May 1980. My Regiment was a 1960s amalgamation of some fine old infantry regiments which fought across the world and garrisoned turbulent parts of the British Empire for over 300 years: the Suffolk, Essex, Norfolk, Leicestershire and Lincolnshire Regiments to name but a few. In early 2002 I found myself commanding my Regiment in Kabul in the wake of 9/11. In 1842, just 160 years before our post 9/11 arrival, our forebears of the 44th of Foot, the Essex Regiment, had been completely destroyed just outside of Kabul by Afghan tribesmen. Each of these 1840's soldiers were equipped with a trusty smoothbore musket that was often hopelessly outranged by the more accurate Afghan jezail, and was completely unsuited to the rocky, broken terrain of central Afghanistan. By the time the British Army returned to Kabul in their victorious campaign of 1879 those Infantry soldiers had been re-equipped with a new rifle-barrelled weapon that allowed them to dominate the battlefield as never before. The seemingly simple technological breakthrough of adding spiral grooves to the inside of the musket heralded a revolution in the primary professional skills and tactics of the infantryman. This 19th century revolution in weapons technology is still central to the Infantry Profession of Arms today.

At the Royal Military Academy Sandhurst in 1980 I was trained with rifle and bayonet as the primary tools of my profession. Just like the soldiers of the Essex Regiment in 1842, I was taught that my rifle was more important to me than anything else, and my skill with it would tip the balance between victory and defeat in battle. But unlike my ancestors whose lethality with the musket depended on delivering massed and often wildly inaccurate volleys at very close range, our rifles allowed us to mass fire at the limit of our vision and hit individual targets with great precision at astonishing ranges. Rifle technology together with skill in marksmanship now allows a skilled sniper to kill at ranges out to 2 kilometers, and platoon firepower delivers a degree of lethality that Napoleonic divisions could only dream of. Rifle technology in terms of materials, automation, ammunition and sights still evolves at an amazing rate. And yet, at the heart of this step change from muskets that used a metal tube to channel an explosively propelled slug towards the enemy over short distances, it is the rifling in the barrel that still makes the all the difference. As a consequence of this technological advance, despite infantry combat still being centred on foot soldiers using personal courage and physical stamina to close with and kill the enemy, land warfare in general has changed out of all recognition.

As the brilliant historical analysis in this book sets out, this was not an over-night revolution in all aspects of the infantry battle. As with the much later advent of the armoured warfare in World War One, it is the early years of adoption of new technology that give rise to the intellectual and institutional change that is so fascinating to study and draw lessons from. Adoption of new technology into armies is never straightforward or uncontested; it is often messy, controversial and full of unforeseen and unwanted effects, and armies in peacetime can be extraordinarily cautious when it comes to change. It is war and the extreme imperatives of battle that drive change into and across institutions most quickly. *The Destroying Angel* charts in great detail the early adoption and employment of era-defining rifle technology across a number of armies, as forged by the fire and battle of military campaigns around the world. It is a fascinating and too often unstudied period of warfare that heralded the modern armies of today. I commend this book as a reference for this extraordinary period of change in warfare. *The Destroying Angel* brings to life the birth of the modern Infantry.

Lieutenant General (ret.) Phil Jones CB CBE DL
The Royal Anglian Regiment

1
INTRODUCTION

On the early battlefields of the Crimean War, densely packed columns of Russian soldiers attacked thin lines of British infantry on several occasions. With heroic bravery the Russian columns went forward, only to be cut down by effective rifle fire while still hundreds of yards away from the British lines. William Howard Russell, the famous *Times* correspondent, poetically described the hapless Russian infantry being swept away by a new weapon that "cleft the enemy like a Destroying Angel."

That destroying angel was the rifle-musket, and the hapless Russian soldiers who "fell like leaves in autumn" under the firepower of this new weapon were the first to experience the anguish of being on the wrong side of a paradigm shift in modern warfare. We can't blame the Russians for failing to see the coming battlefield revolution. After all, for well over two centuries before the Crimean War, European soldiers had gone into battle carrying the exact same infantry weapon: the smoothbore musket. Battlefield tactics had developed around these smoothbore weapons, with linear formations intended to concentrate and maximize close-range firepower. Individually, the smoothbore musket was very inaccurate and hitting a man-sized target at 150 yards with a smoothbore was, more often than not, the result of luck rather than a feat of skill. When massed into large formations, however, the blast of concentrated musketry could be devastating, especially at close range. By the outbreak of the Crimean War in 1854, nearly every aspect of battlefield tactics had been built around the capabilities (and inherent limitations) of the smoothbore musket. When the Russian troops rose up and advanced in their dense Napoleonic columns towards the distant British lines, they had no reason to expect anything different. Instead, they marched into a paradigm shift, and learned firsthand the crushing consequences of being left behind. They died in heaps, and today their tarnished buttons and badges on long-forgotten Crimean battlefields are dug up by relic hunters, who sell them cheaply on the internet to Western militaria collectors.

Opposing these Russian soldiers in the Crimea were the allied armies of the United Kingdom, France, and the Ottoman Empire. The British troops had recently been issued a new weapon, although externally it looked no different from their previous muskets and they continued to march in the same ordinary tactical formations developed for use with the smoothbore. While the British infantry did not *appear* any different, at the battles of the Alma, Balaclava, Inkerman, and at the long siege at Sevastopol, they opened accurate fire at ranges three or four times beyond the range of the Russian troops. More importantly, this long range fire was enormously destructive. To the astonishment of the Russians, the majority of the British troops were not armed with the simple old musket, but rather a modern weapon. This was the rifle. Four grooves, slowly twisting in the barrel of the Pattern 1851 rifle-musket, imparted a stabilizing spin to the bullet that delivered it to where the soldier was aiming. For the first time since the firearm had been adopted for military service, the common infantryman could now individually aim his weapon and strike his enemy at ranges that are still considered more than respectable today. The brilliant Russian engineer Eduard Todleben summarized the situation well: "Our muskets could not reach the enemy at greater than 300 paces, while [the British] fired on us at 1,200."[1] William Howard Russell (whose dramatic prose inspired Tennyson's immortal poem, the *Charge of the Light Brigade*) described the new British rifle as "the king of weapons."

The rifle-musket first entered military service in the 1840s, and was adopted by nearly every Western army by 1855. Scarcely a decade later, it was wholly obsolescent by 1865 as breechloading technology became practical. Quickly abandoned for more

---

[1] Eduard Todleben, Описание обороны г. Севастополя: Частъ I (St Petersburg, 1863), 194.

modern weapons as firearms technology improved with dizzying speed in the second half of the 19th century, the rifle-musket's legacy is anything but clear. The effectiveness of the rifle-musket, and the impact it caused, is vigorously debated among historians; of particular controversy is the rifle-musket's performance in the American Civil War. Was it a decisive and transformative modern infantry weapon that resulted in permanent changes to training, tactics, and strategy? Or did it fail to cause dramatic changes due to inherent limitations and difficulty of use, as American Civil War historians Paddy Griffith, Dr. Earl Hess, Dr. Allen Guelzo, and others have recently argued?

The rifle-musket was, in fact, *both*: in the hands of trained soldiers, in an army that valued instruction and had urgent contextual reasons to promote rifle shooting across the total force, the rifle-musket abolished many prior conventions of battlefield tactics and was decisive in combat. Like all modern infantry weapons, which require the soldier to be systematically trained in the use of a weapon system, the rifle-musket was the first modern infantry weapon carried by professional soldiers. It forever changed the status of the soldier, transforming him from an expendable automaton into a skilled, educated artisan. The rifle-musket also left its mark on tactics and operations, by making it possible for infantry to win on the battlefield by relying solely on the firepower of their weapons, instead of closing with the bayonet.

On the other hand, when issued to enthusiastic yet wholly untrained volunteers led by equally enthusiastic yet untrained leaders (as the case was in the American Civil War), it should not surprise us to find that the rifle-musket's inherent capabilities could not be fully used in any large scale. Understanding the impact of the rifle-musket in the varying contexts in which it was historically used -- both by professional trained armies and by untrained and inexperienced volunteers -- is the objective of this work. A comprehensive approach to the rifle-musket, considering the full breadth of wildly varying contexts in which it was used, provides ample evidence that the rifle-musket, while primitive in many aspects, was still the first modern infantry weapon.

What *is* a "modern infantry weapon," after all? Military historians have proposed countless types of weapons, usually repeating firearms, as the first "modern" weapon. Similarly, historians have argued over the first modern or industrial war. My criteria for the first modern weapon is simple: the first modern infantry weapon was that which has more in common with the weapons carried by soldiers today than the arms that came before it. The standard infantry weapon today (for the U.S. Army) is the M16 rifle, which is rapidly being replaced by the smaller yet essentially identical derivative, the M4 rifle. At first glance, it may seem absurd to compare a gas-operated select-fire magazine-fed centerfire carbine to the muzzleloading, single-shot, percussion rifle-musket. Even so, it is also important to consider that the technological features of the M16 -- gas-operated, magazine fed, etc. -- are only mechanical functions that automatically reload the weapon and reduce the number of operations the soldier must perform in order to fire it. They do not change the fundamental feature of the M16-type weapon that makes it an effective modern arm: it is a *rifle*. Grooves in the M16 barrel impart a stabilizing twist on the bullet with exactly the same principle that stabilizes the heavy lead cylindrical ball fired from the rifle-musket. These modern refinements only enhance the usefulness of the rifle, but without the rifled barrel, they would offer relatively little advantage. Consider the uselessness, in modern infantry tactics, of a smoothbore M16 that wildly throws bullets in the general direction of the enemy. Because rifling grooves allow for accurate shooting, the soldier can aim his weapon – be it a P1853 Enfield rifle-musket or an M4 automatic carbine – and individually engage targets out to several hundred yards. To effectively use such a weapon of precision, soldiers must be specifically *trained* in their use. The rifle-musket meets all these criteria, while the preceding weapon type, the smoothbore musket, does not.

The American Civil War, by virtue of being by far the largest and bloodiest war of the rifle-musket era, has typically been the only contextual focus for evaluating the rifle-musket's contribution to military history. This is, unfortunately, a common but flawed approach, that inexplicably ignores the highly successful employment of the rifle-musket in broad and varied contexts, in other armies and conflicts. In 1854, seven years before the Civil War, British troops with rifle-muskets demonstrated the massive increase in the power of regular infantry and forced a wholesale reconsideration of the conventional battlefield deployment of cavalry and artillery. Three years later, the effectiveness of trained soldiers using the rifle-musket was demonstrated in numerous engagements during the Indian Rebellion of 1857. Four years before the bombardment of Fort Sumter, British officers at the dedicated School of Musketry were theorizing about embryonic forms of fire and maneuver warfare, using elements of troops with rifle-muskets to provide cover and suppressing fire for friendly attacking elements. The 1859 *Field Exercise and Evolutions of Infantry* manual embraced the new fire tactics and described soldiers who could not shoot as "useless" and an encumbrance to the unit. By the time American armies of poorly trained volunteers were slaughtering each other at close ranges at Bull Run, Antietam, and Fredericksburg, thousands of British officers and NCOs had attended the School of Musketry and thorough musketry instruction for every soldier in the army had been a mandated by the Queen's Regulations for nearly a decade. The goal was an entire army (and even a large corps of Militia and Rifle Volunteers to supplement the army) trained to shoot, with the knowledge and skill necessary to employ the rifle-musket to the full extent of its capabilities.

In the hands of trained soldiers, serving in the army of a nation that devoted the finances and resources for comprehensive instruction and live fire training, the rifle-musket was extremely effective. Trained soldiers compensated for the parabolic trajectory of the slow-moving bullet by elevating their barrels, and delivered fire (in documented combat conditions) with effect at ranges beyond 400 yards. It was unmistakably decisive in the Crimean War and the Indian Rebellion. It shattered Napoleonic convention almost overnight; prescient theorists and tacticians realized dense columns and close-order battle lines were relics of the smoothbore age. Future battle would be in open order, and soldiers would utilize cover and concealment to a far greater extent than earlier skirmishers. Infantry power increased dramatically, forcing a reconsideration of the role, use, and armament of cavalry and artillery. It was, simply, a rifle in the fullest modern sense. While still hobbled by the limitations of muzzleloading and black powder propellant, it is the direct ancestor of the rifles that remain the individual soldier's weapon of today. No smoothbore weapon could have had this impact, even if the rate of fire of such a smoothbore weapon could have been effectively increased.

Yet the rifle-musket was even more than just a particularly revolutionary arm that, like so many others, changed how warfare was conducted. It was the first modern infantry weapon, and was correspondingly carried by the first modern soldier. Just as the rifle-musket has more in common with the modern rifle than with the old smoothbore, the trained soldier who carried the rifle-musket has more in common with the infantryman of today than with the Napoleonic soldier. This work focuses primarily on this period of great transition in the British Army, and not simply out of the convenience of a common language and shared Anglo-American military tradition. We focus on the British Army simply because they turned the corner first, and embraced *firepower* as the future of infantry combat before any other world military. All of the major powers in the 1850s understood that the rifle-musket required specialized training to be used effectively; the British Army was the first to plow through intense controversy and entrenched resistance to the new technology, commit to the institutional changes necessary and shoulder the vast expense of implementing this transition. Instead of merely contributing one random musket ball in a massive volley, the British soldier was now expected to skillfully judge the distance to his target, aim correctly, and exercise sufficient marksmanship fundamentals to hit what he aimed at. After about 1859, the British Army anticipated winning future battles primarily with infantry firepower; nearly all other powers of the rifle-musket era retained the bayonet as the primary infantry weapon and put their faith in Napoleonic shock tactics.

Not only did the rifle-musket increase the power of infantry as a branch, but it increased the importance and social standing of the common soldier. No longer was the soldier a mere ignorant automaton, drilled at movement in rigid formations until he could do it without thinking, and responding only to the specific orders of his officers. To effectively use a rifle, the soldier had to think, he had to be trained in the use of a complex weapon system, and he had to be trusted to operate with some degree of independence on the battlefield. This transformation of the common soldier from Napoleonic automaton to a trained, skilled rifleman is just as significant (if not more so) than the transformation of the smoothbore musket to the rifle. We see, in the British Army soldier of the mid-19th century with an Enfield rifle in his hands, the embryonic forms of the modern soldier's essential qualities.

2
THE INTRODUCTION OF THE RIFLE-MUSKET

From about 1450 to 1850, a period of about four centuries, military firearms changed surprisingly little. The improvements made during this period were to the quality of the gunpowder and the methods of igniting the gunpowder charge that propelled the bullet (almost universally a round lead ball) out of the barrel. Throughout these centuries, the reliability of firearms improved as the matchlock was replaced first by flintlock and, starting in the 1830s, the percussion lock. Gunpowder was improved upon considerably, being pressed, corned, and glazed before sorting by grain size. By the 1850s, the powder produced at Waltham Abbey for the British Government was essentially perfect and even superior, in virtually all respects, to modern commercially-produced black powder used today for sporting purposes (primarily due to the higher quality of the charcoal used).

The gun barrel, however, remained simply an unadorned iron tube. Military arms were loaded by ramming a generously-undersized ball down the barrel, still wrapped in the cartridge paper; this was to ensure the ball did not simply roll out of the barrel if it was tipped below the horizon. When fired, the ball flew blindly down the barrel, often bouncing a couple times off the interior of the barrel as it went, and its trajectory was unpredictably wild. Colonel Ernest C. Wilford, the chief instructor at the British Army's School of Musketry in the late 1850s, described the odds of being hit by an aimed smoothbore musket ball at 300 yards as about the same proportion "as a farthing to the National Debt." He also declared, tongue in cheek, to be willing to stand and be fired at all day long by a smoothbore musket at the range of 300 yards, provided the shooter promised to faithfully aim each shot directly at him. These anecdotes illustrate the inaccurate nature of the smoothbore musket at ranges beyond 100 yards. Colonel Wilford could remember the days as a junior officer when "Brown Bess" armed the British Army, and soldiers closed their eyes, looked away, and threw open their mouths (to reduce the felt effects of concussion) when firing these muskets. Yet a good soldier could fire up to four rounds per minute with a smoothbore musket, as the loose-fitting ball allowed for rapid, easy reloading even during periods of sustained fire. Wellington and other British veterans of the Napoleonic Wars praised Brown Bess and credited the speed, discipline, and crushing effect of British musketry for victory at Waterloo and countless other engagements. By 1850, the power of British musketry had become almost legendary; to speak of replacing Brown Bess with any other kind of weapon was a sort of military heresy.

Before the 1840s, rifles were only employed on the periphery of battle. While their use for hunting and sport was fairly common, the need for the bullet to fit tightly in the rifled barrel in order to receive the stabilizing spin meant that rifles took much longer than a smoothbore musket to load. Rifles were appreciated for their greater accuracy, but the usefulness of a weapon that sometimes took several minutes to reload was debatable, particularly on a battlefield where the ordinary soldier could fire four to six rounds in the time it took a rifleman to load and fire once. Napoleon Bonaparte, the master of smoothbore musket military tactics and strategy that defined an entire era of warfare named after him, did not consider rifles to be worth the expense, slow rate of fire, and extensive training that the soldier required before he could use the weapon to its potential. The ranges at which infantry exchanged fire in this period rarely exceeded 100 yards. Tactics and strategy developed around this weapon system, or more specifically, the inherent limitations and roll-of-the-dice accuracy of the weapon system.

In the 1830s and 1840s, several French officers (among them Delvigne, de Thouvenin, and Minié) incrementally developed a new type of bullet that allowed a rifle to be loaded as quickly as the smoothbore musket.[2] The most refined of these, the Minié bullet, was smaller than the size of the rifle barrel by a few thousandths of an inch, facilitating fast and easy ramming. When the rifle was fired, the bullet was expanded to fill the grooves of the rifling by the pressure of the ignited powder charge. Minié's bullet, and a rifle to fire it (the French Minié rifle), was adopted by the French Army in limited numbers, but its capabilities caused something of a panic in the United Kingdom. The British press wildly stoked fears of a French invasion led by Minié rifle-armed troops that would outclass British troops armed with smoothbore percussion muskets. In astonishingly short order, the British essentially copied the French rifle and adopted it as the Pattern 1851 rifle. It came to be popularly known, fittingly, as the Minié rifle. Even though it still fired a massive .690-caliber bullet that weighed well over an ounce, the rifle was fitted with sights ranged out to 900 yards. To the astonishment of military authorities, hitting generously-sized targets (such as a group of artillery pieces, or a formation of cavalry) at 900 yards with the P1851 rifle was by no means difficult for a trained shooter.

The impact of the adoption of a rifle as a general-issue weapon for the army cannot be overstated. British troops, bound for the Crimean War, left England while still mostly armed with the P1842 (or older P1839) smoothbore musket, a weapon with a maximum effective range of about 150 yards and (except for its improved percussion lock) still essentially the descendent of the ancient arquebus. They brought a few of the new rifles with them, and a few more arrived while at Malta. In the summer of 1854, while disembarked at Varna, large shipments of the new P1851 Minié rifle arrived, and armed most of the divisions of the British infantry force destined for the battlefields of Crimea. Instead of being just another descendant of the old smoothbore musket, the P1851 was the first of the modern infantry rifles to see extensive combat service. For the first time, the rifle was no longer

---

[2] Two Englishmen, Greener and Norton, also developed self-expanding bullets but their work was spurned by military authorities. The concept found much more fertile ground in France, though some decades later.

relegated to the periphery of battle. At Varna, the British soldier turned in his smoothbore musket that belonged in the antiquated era of bayonet-centric Napoleonic shock tactics, and he picked up a modern infantry weapon.

The Pattern 1851 rifle, by virtue of its capabilities and the increased power it gave the infantry soldier, had more in common with the S85 or M16 rifles than with the smoothbore musket the soldiers had turned in to storage. With the Minié rifle, the trained individual soldier could now accurately engage individual targets out to approximately 400 yards, and area targets to 800 yards and beyond. The troops were familiarized with the new rifle at the firing ranges on Malta, and again at Varna; the astonishing capabilities of the new weapon were not lost on the commanders or the men.[3] Even before the first British troops landed on the Crimean Peninsula, a particularly astute sergeant major writing in 1854 observed that the new Minié rifle was "a weapon destined to exercise a powerful influence on Military pursuits and operations."[4] He was correct indeed, but perhaps too prudent; the new rifle was going to turn the world of military pursuits upside down.

By the outbreak of the Crimean War in 1854, European warfare had been codified into rules and conventions commonly (but perhaps too simplistically) known as Napoleonic warfare. Land warfare was conducted with three branches: infantry, artillery, and cavalry. The infantry was overwhelmingly the most numerous of the branches that held the field, conducting the bulk of the fighting and finding strength only in massed formations.

The artillery provided long range fire support to assist the infantry, and cavalry supported by scouting, and in rare or ideal circumstances, decisively charging. Infantry, armed with the smoothbore musket, were unable to effectively engage when enemies were beyond about 200 yards and were highly vulnerable

---

[3] *The Thin Red Line: The Regimental Paper of the 2d Batt., Princess Louise's, Argyll & Sutherland Highlanders*, Volume 10, Number 8, December, 1904, 123

[4] Thwaites, J., *Observations on the Minié Rifle* (W Clowes and Son: London, 1854), 2

if exposed, alone and unsupported, to enemy artillery or cavalry. Wilkinson, the British gunmaker and inventor, wrote in 1852 that "it would be difficult to imagine anything more inefficient than the fire of English [smoothbore] musketry at any distance beyond 200 yards."[5] There was very little that unsupported infantry could do against artillery firing at them from long ranges. Military convention required unsupported infantry to "form square" if charged by cavalry, since with smoothbore muskets the infantry would only have time to fire a single volley before the cavalry crashed into them. The powerful charge of cavalry against disorganized or breaking infantry was the holy grail of battle and the seal of victory, pursued by commanders even into the 20th century.[6]

These entrenched conventions resisted the impending change brought by the Minié rifle. The *Aide-mémoire to the Military Sciences*, published in 1850, defended the traditional role of the line infantry with smoothbore muskets, firing at under 200 yards and relying upon the bayonet, rather than bullets, for decisive action. The writer passionately condemned the "erroneous ideas" of the rifle's advocates:

> Troops on service do not halt to play at long bowls: a field of battle presents a series of movements for the purpose of out-flanking or closing-in upon your adversary, and when within 200 yards, to deliver your fire with effect. Firing at 500 or 600 yards is a business of artillery and therefore when infantry fire is given at these long distances, or even 300 or 400 yards, it is a misapplication of the musket, a loss of time, and tends to make men unsteady; it is besides a waste of ammunition.

---

[5] Henry Wilkinson, *Observations - Theoretical And Practical - On Muskets, Rifles And Projectiles*, (London: 1852), 2
[6] Vast numbers of cavalry were held in reserve on the Western Front of the First World War by both sides for some years, awaiting the "breakthrough" that never came.

Such was the military convention in 1850, and lest we balk at the stubborn intransigence of military authorities to accept change, we do well to remember that the smoothbore musket had scarcely seen any improvement in the preceding 200 years. Men were born, served their time in the army, retired, and in their old age might observe their grandsons carrying virtually the exact same musket. Over these 200 years any number of new weapons had been proposed, from breechloaders to air guns, and none could dethrone Brown Bess (or her slightly more reliable percussion derivatives). An article in the *London Quarterly*, in April 1852, suggested that Brown Bess should simply be "thoroughly overhauled" by reducing the bore and simplifying construction, and to replace Brown Bess "only when you are quite sure of your substitute."[7] The debate would continue, and even though the advocates of the rifle were ultimately successful in their arguments, there were still those in the highest positions of military authority calling for the return to the smoothbore as late as 1857.[8]

In Dr. Christopher Roads's wonderful *The British Soldier's Firearm*, the adoption of the Pattern 1851 rifle is covered in concise detail; I cannot recommend this work too highly for anyone seriously interested in the specifics of British military arms 1850-1864. For our purposes, however, it is sufficient to note that the adoption of the P1851 was approved by the elderly Wellington in spite of the Iron Duke's suspicion of rifles. When Wellington died in 1852, he was succeeded by the Viscount Hardinge, a more forward-looking advocate of rifles. It was Hardinge and his allies who, against much objection, ensured that Pattern 1851 rifle was rapidly produced in significant numbers (although there was endless trouble with the various contract

---

[7] *London Quarterly, Vol. 90* (London: John Murray, Albemarle Street, April 1852), 446.
[8] *Hansard's Parliamentary Debates, Third Series, Commencing with the Accession of William IV, Vol CC,* 1870 (London: Cornelius Buck, 1870), 2063

makers) and, ultimately, sent to arm the British troops bound for war in the Crimea.[9]

In March 1854, Lord Hardinge urgently dispatched the crates of brand new Pattern 1851 rifles to Malta, where the British Army regiments were gathering. The crates were accompanied by Captain Augustus Henry Lane-Fox, one of the earliest advocates of the rifle for the general weapon of the British Army.[10] He was appointed among the first instructors at the newly opened School of Musketry at Hythe in 1852. He was also the perfect man for the task of providing musketry instruction to the British troops bound for the Crimea. Lane-Fox wrote most of the *Instruction of Musketry* manual, published in February 1854.[11] The crates of Pattern 1851 rifles and Captain Lane-Fox arrived at Malta as British troopships were bringing regiment after regiment. Lord Hardinge "refused to allow under-instructed men to proceed to the Crimea," and directed Lane-Fox to open an improvised musketry school at Malta.[12]

The units at Malta commenced an urgent immersion in musketry instruction. A regimental official history recalls that "while at Malta the men had constant target practice with the Minié rifle."[13] Under the recently-published *Instruction of Musketry*, officially every new soldier was to fire 110 rounds in musketry instruction and qualification. It's unknown how many rounds the

---

[9] For a fuller discussion of the Pattern 1851 and its ammunition, see Brett Gibbons, *Pattern 1853 Enfield Ammunition: A history of the ammunition used in the first successful general-issue military rifle,* 2016

[10] Lane-Fox inherited his cousin's estate, took the name Pitt Rivers, and is well known today for his work in archaeology; his role in military history with the adoption of the rifle is much less known.

[11] Bowden, Mark, *Pitt Rivers: The Life and Archaeological Work of Lieutenant-General Augustus Henry Lane Fox Pitt Rivers* (Cambridge University Press: 1991), 15.

[12] Strachan, Hew, From Waterloo to Balaclava: Tactics, Technology, and the British Army 1815-1854 (Cambridge University Press: 1985), 50.

[13] Burgoyne, Roderick Hamilton, ed., *Historical Records of the 93rd Sutherland Highlanders* (Richard Bently and Son: London, 1883), 110

soldiers at Malta were permitted to fire, but it may be assumed from the period sources that it was a considerable quantity. There were not enough rifles for all the British troops, so they were distributed to the best shots of the companies. All soldiers were familiarized with the rifle, and not just those who had been issued the new weapon. After several weeks, the British forces began moving to the Black Sea, staging at Scutari and then primarily at Varna, on the Bulgarian coast. More Pattern 1851 rifles arrived in number sufficient to arm all but one of the British infantry divisions. "We were trained to the Minié rifle before the Regiment embarked [from Malta], and at all the time the Regiment lay at Bulgaria and at Varna," a veteran of the 93rd Highlanders wrote later.[14]

Finally, in September 1854, the combined allied forces of Britain, France, and the Ottoman Empire proceeded from Varna to the Crimean Peninsula. They landed on the 14th, and while the French had come prepared with all manner of logistical necessities, the British troops had no tents or supplies whatsoever except what they carried ashore on their persons. Officers and men alike suffered at the landing point, drenched by two days of heavy rain. It was the first of countless, persistent logistical and medical shortcomings that would eventually define the Crimean War (along with the Charge of the Light Brigade) as a clumsy, bungled affair. But the British soldiers who lay soaked to the bone near the brackish Lake Touzla cradled their Pattern 1851 Minié rifles, sealed against the wet by pump grease and a muzzle-stopper to keep the rust away from the four slow spiral grooves cut inside the barrel. They didn't know it, but these miserable soldiers were about to fundamentally change the nature of Western warfare in the coming months. Even so, tragically few would ever return safe and healthy to Britain.

The Minié rifle was used to devastating effect at the Battle of Alma, the first major battle on the Crimean Peninsula. It was the first large-scale use of rifled infantry weapons in history. Prior to

---

[14] *The Thin Red Line: The Regimental Paper of the 2d Batt., Princess Louise's, Argyll & Sutherland Highlanders,* Volume 10, Number 8, December, 1904, 123

the Alma, rifles were employed on the periphery of battle by specialized companies of picked soldiers. On 20 September 1854, more than two thirds of the British regular line infantry engaged (and a smaller percentage of French) carried the Minié rifle. As the first Crimean battle of the war, with the novelty of the new rifle highly appreciated by period commentators, the Alma (and the rifle-musket's role) has been studied more extensively than subsequent battles prior to the American Civil War.

The battle was an attack against strong Russian positions, with earthwork fortifications on hills overlooking the river. While British (and French) losses were significant, the rifle was decisive; Alma is one of few 19th century battles where the attacking force suffered fewer casualties than the defenders. Some British infantry regiments advanced while firing, a tactical innovation done on the spot. French troops, armed with a Minié rifle, engaged Russian artillery with infantry rifles at long range, another tactical first. One Russian battery of 100 men lost 48 men and most of their horses to Minié rifle fire at "more than 900 yards" and "prevented [them] coming close enough to deliver their fire with effect."[15] The Russians responded with an attack in column, a typical Napoleonic procedure; the columns were destroyed by rifle fire. Two Russian regiments in column, the 31st and 33rd Regiments, converged on a lone Grenadier battalion, unsupported and separated from the rest of the allied lines. Orders were sent to "retire the Guards" but the order was not received, and the Grenadiers held their ground. Under the old rules of warfare, these dense columns should have easily and decisively driven a lone battalion back. Instead, the thin British line, "immovable as a rock," formed in two ranks, maneuvered to fire into the front and flank of the Russian columns. The range was not mentioned, but since it was recalled that a soldier asked his officer "to what distance he should set the sight of his Minié," we can safely assume it was well beyond the old smoothbore

---

[15] Russell, William Howard, General Todleben's History of the Defence of Sebastopol, 1854-5: A Review (Tinsley Bros, London: 1865), 43

range.[16] The Russian columns were shot to pieces, and the shattered remnants fell back in disorder. Stunned and impressed by the battlefield debut of the rifle-musket, the Russians learned quickly from the bloody lesson.

---

[16] Hamilton, Frederick William, The Origin and History of the First or Grenadier Guards (London: J. Murray, 1874), 192

3
THE DAY BATTLE CHANGED: OCTOBER 25, 1854

A paradigm shift in battlefield tactics occurred on October 25, 1854, when ordinary British rifle-armed infantry cut down a Russian cavalry charge and silenced Russian artillery at the Battle of Balaclava. It is ironic that this date, and the battle itself, is far better known for Lord Cardigan's disastrous charge of the Light Brigade at the very end of the engagement. Alfred, Lord Tennyson's poem *Charge of the Light Brigade* ensured that Balaclava would forever be associated with someone's blunder, and soldiers of a bygone era who were not to reason why, but to merely do and die. The charge of the Light Brigade eclipsed the decisive demonstration of the power of rifle-armed infantry over cavalry and artillery in this chaotic Russian assault upon the weak allied lines defending the port of Balaclava. Yet in this battle, the rifle-armed infantry soldier in the 93rd Highlanders "thin red line" stood closer in the military paradigm to the current modern soldier than the musket-armed Napoleonic soldier he used to be.

The British lines defending Balaclava were weakly defended, with six crude redoubts manned primarily by third-rate Turkish recruits. They were not particularly a high priority for supply distribution either, and many of the Turks had little or no ammunition, food, or even water. Yet they fought remarkably hard when the Russians attacked, even though when they finally broke, the panic was contagious and the Turkish defensive line collapsed. Had the Russians decisively committed, and pressed a strong attack without hesitation, they might have achieved a substantial victory. But this was open ground, and the Russian commanders were justified in fearing what would happen if their smoothbore musket-armed troops were caught in the open field under fire by British troops with the Minié rifle. They had suffered terribly from the Minié rifle at Alma. But the Russian cavalry were not as gunshy. Maj.-Gen. Sir Henry Hugh Clifford, VC, recalls in his diary that, on October 3, 1854, his dinner was interrupted by a report of "300 or 400" Russian cavalry approaching within 1,000 yards of a 40-man outpost. Clifford went out and found the captain in charge of the outpost "in a great stew." The cavalry had by this point already moved away, but Clifford chided the captain. "I was sorry to see he had not more confidence in his men, his arms and his position," Clifford wrote in his journal. Even if the Russians had *double* the number of cavalry, "with 40 Minié rifles opening fire" by the time the cavalry could cross the 1,000 yards of ground, "they should be so disorganised and cut up that no danger whatsoever was to be feared." The general wrote this off to inexperience. "A great number of our men are so young and know so little about the Minié rifle and the wonderful distance it carries (1,000 yards), that many opportunities of this kind are lost."[17]

Content with holding the redoubts captured from the Turks (and hauling away British guns that had been assigned to the Turkish redoubts), the Russian infantry prudently chose not to advance on the final line of defense, which included the distinctive 93rd Highland Regiment in kilts and feather bonnets, formed up into line with Turkish troops. The Turks had smoothbore muskets and were already weary from hard fighting, and were probably low on ammunition; they would ultimately flee. Behind the 93rd was Balaclava, easily visible from the redoubts the Russians had captured from the Turks. But the Russian infantry would not advance, knowing that while they slowly crossed the thousand yards of open ground between them and the British troops, they would be receiving effective rifle fire the entire way.

The 93rd was, however, now within range of Russian artillery, which opened an accurate and effective fire on the Highland regiment. Sir Colin Campbell, the commander of the 93rd, responded by pulling his troops back slightly and having them lie down beneath the protection of a reverse slope, where the Russian artillery fire could not hit them. Russian cavalry

---

[17] Clifford, Major General Henry Hugh, V.C., *Henry Clifford V.C., His Letters and Sketches from the Crimea* (Pickle Partners Publishing: 2016), 58

seized this opportunity, and a "large body" of cavalry (the exact number of cavalry is debated) began advancing rapidly towards the 93rd. Under the traditional conventions of Napoleonic warfare, Campbell and the 93rd Highlanders were in serious trouble. If they stood and formed a tightly-packed square to resist the cavalry charge, as doctrine stipulated, they would be a dense and glorious target for the Russian artillery. The effect of artillery round shot upon tightly formed infantry squares was carnage almost indescribable. Campbell could keep his troops in line, but the line formation was thin and exceedingly vulnerable to being broken and totally overrun by a charging cavalry. Essentially, the old rules of warfare required 93rd Highlanders to retreat or be destroyed, either by bouncing round shot tearing through their squares formed to guard against cavalry, or by masses of Russian cavalry smashing through an easily broken thin line formation. There was no value in retreat, as Balaclava and the sea was to the backs of the Highlanders. When cavalry charged, they would slowly increase the gait of their horses until, at the very last, they would break into the full gallop that would cover 100 yards -- the effective range of the smoothbore musket -- in mere seconds. With the old musket, infantry being charged by cavalry had all of a few seconds to fire effectively at the awesome spectacle of the approaching enemy, before the weight of horse, man, and steel hit home. And this final volley, delivered just before cavalry crashed into a fragile line of flesh and bone, would invariably be a shaky and uncertain fire, even with the most disciplined of soldiers. The square, then, was the only recourse for infantry facing cavalry, as the horse instinctively would refuse to charge into a dense mass and, instead, would turn to one side or the other of the square. To not form square, and remain in line, was madness under the old principles of war.

    Campbell knew this, and his soldiers knew this as well. Even so, he kept the 93rd in line, advanced them to the top of the slope, and rode along the line shouting, "Men, remember there is no retreat from here. You must die where you stand." He did not form square, in direct contravention of every tactical rule for infantry facing a cavalry charge. In a square, three out of four sides would be unable to fire at the approaching cavalry. Campbell put his faith in the rifle. "He knew his weapons, and he knew the men who wielded them," the gunmaker and advocate of the volunteer movement, James Dalziel Dougall, wrote of Campbell in 1859. "This was the greatest of all heroism -- it was moral and physical courage combined, and both resting upon science."[18] William Howard Russell immortalized the line formation of the 93rd with his description of a "thin red streak tipped with steel," but in popular memory it has gone down in history as the simple "thin red line."

Most eyewitness accounts of the engagement agree that Campbell fired the first volley at about 600 yards, "when Sir Colin thought that our Minié rifles might reach the enemy."[19] At this point, the Russian cavalry would have been moving at a quick trot. This was the least effective of the three volleys that the 93rd ultimately fired, owing to the range, but it must certainly have unnerved the Russians to hear the unexpected rotary hiss of rifle bullets flying about them. Only a few bullets from this first volley actually hit horse or rider. The Russians kept on, and increased their speed. The 93rd reloaded. At the distance of 350 yards, Campbell fired his second volley. When the smoke cleared, from the British line the volley did not seem to have had much effect, but a Russian officer later explained that "we received your second volley, by which almost every man and horse in our ranks was wounded."[20] Such a destructive volley, at 350 yards, would have been utterly impossible with smoothbore muskets. This shook the Russians, and as they began to incline to their left, intending to wheel around the right of the British line, the third volley was fired at about 150 yards into the Russian flank. Instead of wheeling around the 93rd, the remaining Russian cavalry

---

[18] Dougall, James Dalziel, *The Rifle Simplified* (Thomas Murray & Son: London, 1859), 40.
[19] Munro, William, *Reminiscences of Military Service with the 93rd Sutherland Highlanders* (Hurst & Blackett: London, 1883), 36.
[20] Ibid., 39.

turned all the way round, and withdrew back in the direction they had come from. Campbell, with his characteristic admirable chivalry, did not allow his soldiers to fire into the surviving Russians as they retreated. The effectiveness of the volleys would be questioned later because only a few Russian horses and cavalrymen were left on the field; the Russian officer who survived the charge explained that many wounded men managed to cling to their horses in the retreat back to their lines. To fall from the saddle would mean inevitable death lying exposed and untreated on an open field before the enemy's forces.[21]

And thus did a thin red line of infantry decisively halt a powerful cavalry charge. Never again on the battlefield would cavalry possess the awesome power that it once had. The glorious cavalry, usually the prized branch of any European army, had been rendered tactically impotent in its most decisive role by the lowly common infantryman. This was studiously noted by many military commentators. "The whole tactics of war are already changed," Dougall wrote in 1859. "With the new rifle, cavalry on level ground will be under fire during the whole charge."[22] It took about eight minutes for attacking cavalry to cover a distance of 1000 yards. The first 400 yards, at a quick walk, took five minutes. The next 400 yards was at a quick trot, covered in two minutes. In the final 200 yards, the horse was brought increasingly to the full gallop and the last hundred yards could, with light cavalry on well-bred horses, be covered in six seconds. This was death for smoothbore-armed infantry without support. It was death for the cavalry when infantry were armed with the rifle. As Dougall observed, cavalry were under the power of infantry for the entire duration of their 1000-yard, 8-minute charge. Competent soldiers could fire 16 rounds in 8 minutes, while the best soldiers could approach 24. Even small company-sized units, therefore, could discharge 2400 rifle bullets before cavalry reached them.

The supremacy of infantry over cavalry quickly percolated through the army. "Don't talk of forming square to receive

---

[21] Ibid., 40
[22] Dougall, 42.

cavalry," Colonel Earnest Christian Wilford of the School of Musketry told his students in 1859, emphasizing that "a taught soldier cannot miss him."[23] Colonel Wilford advocated firing at 250 yards because at this range, owing to the trajectory of the P1853 Enfield rifle then in use, the bullet would never go any higher in its flight than the height of a horse and rider. This meant that even if the distance was improperly estimated, the soldier was still guaranteed to hit as long as his aim was generally on target; this was an impossibility with the old smoothbore. Published while the Crimean War was still ongoing, Captain John Le Couteur's *The Rifle: Its Effects on the War* reached an obvious conclusion: "the power of infantry over cavalry is greatly augmented."[24] Other military writers went even further. "These qualities [of the rifle] must, of necessity, give an importance to infantry superior to any they have hitherto attained," wrote Captain T. J. Thackeray in 1858, "and will modify considerably any operations in which they may be hereafter engaged... Cavalry and artillery would thus lose much of their comparative value in the field."[25]

Not even the artillery was safe from the Minié rifle. For hundreds of years, the artillery (the "king of battle") brought the only weaponry to the battlefield that could effectively put rounds on the enemy at ranges over 200 yards. In 1854, field artillery pieces were light, handy weapons that fired solid shot, true shrapnel shell, or canister out of smoothbore barrels. The weight of shot was usually six to twelve pounds, and they were accurate to around 1000 yards. Artillery was far more effective at closer ranges, where gunners could more clearly see the impact and effect of their fire, and make necessary adjustments to fuze lengths, elevation, etc. Infantry under fire from artillery at 600 yards were absolutely helpless; it was only small comfort that field

---

[23] Edwards, Henry, *A volunteer's narrative of the Hythe course of instruction in musketry* (London: 1860), 17.
[24] Le Couteur, John. *The Rifle: Its effects on the War* (London: 1855), 29.
[25] Thackeray, Thomas James, *The soldier's manual of rifle firing, at various distances* (London: 1858), xiii.

artillery's rate of fire was generally slow, perhaps two rounds a minute or less. Infantry with muskets had to be within 200 yards to have any hope of hitting the artillery crews; the artillery, meanwhile, could begin firing canister or "grape" at about 300 yards, turning their cannon into massive shotguns. Nothing on the battlefield was more devastatingly powerful against infantry than a battery of guns, loaded with canister. But at Balaclava on October 25, 1854, while the 93rd Highlanders were giving the Russian cavalry a taste of modern warfare, a single British lieutenant single-handedly took on a battery of several Russian cannons at long range with a Pattern 1851 Minié rifle. The lieutenant won, and for the second time on that single day in the Crimea, established military doctrine was turned upside down.

Even casual military history enthusiasts have heard of Tennyson's *Charge of the Light Brigade,* or the 93rd's Thin Red Line. Few, however, have heard of Lieutenant Arthur W. Godfrey, an energetic young officer of the 1st Battalion, Rifle Brigade. Born in Jersey in 1829, the teenaged "Gentleman Cadet" Godfrey of the Royal Military Academy at Sandhurst purchased a second lieutenant's commission in 1845 and reported to the Rifle Brigade's 1st Battalion. From the age of sixteen, Lieutenant Godfrey was thoroughly immersed in the military application of the rifle. As the descendant of the storied 95th Rifles of Napoleonic War fame, the regiment lost its numerical distinction in the massive downsizing of 1816, but was retained as the Rifle Brigade. In 1852, Lieutenant Godfrey and the Rifle Brigade fought in the Xhosa War of 1852-53 (the "Third Kaffir War"), where he saw intense fighting and was wounded. While some commanding officers equipped their troops with the new Minié rifle at personal expense during this war, the Rifle Brigade was still using the old model Brunswick rifle. They returned to England in early 1854, and finally received the new Pattern 1851 Minié rifles in June. A month later they were ordered urgently to Malta, and then to Constantinople. Although cholera ripped through their encampment, "while in this camp the Riflemen were frequently exercised in the use of the new arm which they had received before their departure from England."[26]

At the Battle of Alma, Lieutenant Godfrey and the 1st Battalion were present but not engaged (the 2nd Battalion was). After Alma, the Rifle Brigade occupied the trenches besieging Sevastopol. Godfrey quickly distinguished himself as an expert and effective rifle officer. As French and British elements advanced for the very dangerous task of opening the trenches (the first step of establishing a siege), the Russians harassed them with artillery and sorties. On October 11, Lieutenant Godfrey was sent forward in command of one of the many skirmishing parties that preceded the engineers and sappers, to protect the men building the trenches. At one point, during a Russian sortie, the outlying skirmishers withdrew due to a misunderstood order that did not reach Lieutenant Godfrey. His lone element remained ahead of the lines, and in the words of William Howard Russell, "maintained the ground with tenacity, and fired into the columns of the enemy with effect."[27]

When the Russians attacked the defenses at Balaclava, the 1st Battalion was ordered out of the trenches and moved, at the double, to the battlefield. They arrived at approximately 10:30 a.m., over an hour after the 93rd Highlanders had held the thin red line but just in time to observe Lord Cardigan's Light Brigade make its immortal charge. The 4th Division was ordered up, with the Riflemen in skirmish formation leading by wings. They took up positions in the field, guarding the approach to Balaclava by the main Sevastopol road. By this time, after the Light Brigade had charged, the battle was essentially over and the Russians were by no means defeated. When the 4th Division advanced, it presented a lovely target of opportunity for the otherwise-idle Russian artillery, upon the ridge. The Russians "brought forward a field battery of six guns" to within field artillery range of the 4th

---

[26] Cope, William H., *The History of the Rifle Brigade (the Prince Consort's Own) Formerly the 95th* (London: 1877), 300.
[27] Russell, William Howard. *The British Expedition to Crimea* (London: 1877), 143.

Division and opened a "very troublesome fire" on the right side of the line.[28] The battery also threatened Turkish troops, who had just reoccupied one of the contested redoubts. Unlike the 1st, 2nd, and 3rd Divisions, the 4th Division had not received the new Minié rifle, and was the only British division still armed with the percussion version of Brown Bess. The Russian guns, between 800 and 1,000 yards away, could fire with impunity at the infantry. This was the ideal situation for field artillery, casually lobbing shot into enemy infantry formations without the slightest fear of receiving any effective fire in return. At this considerable range, beyond about 800 yards, the Russian artillery had nothing to fear from infantry. They would have been carefully laying their guns, taking their time adjusting elevation and deliberately aiming shots so that the balls would pass diagonally through the British line. Skipping a cannon ball at an acute angle through an infantry line would strike several soldiers, and was far more effective than firing squarely at a perpendicular infantry line. This was the scientific application of gunnery in battle in 1854.

A battery of six guns would have approximately 100 men or more assigned. There were five or six men on each piece, to load and ready the priming while the gunner carefully aimed. Every shot had to be aimed, because the gun moved considerably when fired. Several more men attended the ammunition at the limbers, which were usually kept behind the guns so that an explosion of the limber would spare the gun crew. Artillery and engineer officers were the intellectual men of an army. While the cavalry was dashing and magnificent, the artillery was technological and scientific. Frederick II of Prussia famously said "artillery adds dignity to what would otherwise be a vulgar brawl." Artillery was also very expensive; the misunderstood orders that sent the Light Brigade charging into the valley of death had actually intended Lord Cardigan to stop the Russians from hauling away British cannon seized from the captured redoubts. It is almost certain that this six-gun battery of Russian artillery was among the guns "to the right of them" that fired upon the Light Brigade's charge.

---

[28] Cope, 317.

Now the guns swung their muzzles to the west, to fire at the most helpless target of all: infantry.

The Russian gunners did not know it, but the tactical employment of the entire branch of field artillery was about to be challenged by Lieutenant Godfrey and his small team of rifles. General Cathcart, commanding the 4th Division, sent Lieutenant Godfrey forward with "a few men" to "try to silence those guns."[29] The number of men who went forward with Godfrey is unspecified, described once as a "company" but usually as few men, or a band of skirmishers. It was probably no more than a platoon-sized element, and quite possibly even less. Between them and the Russian battery was 800 yards or more of open ground that "afforded no cover." While the Pattern 1851 rifle was sighted out to 900 yards, hitting a man-sized target at anything over 600 yards was as much a matter of luck as anything. The Russian gunners, probably much surprised, shifted their aim to the small group of crazy Englishmen who were brazenly approaching their gun line. Crawling on their stomachs, with Russian shrapnel and ball roaring just overhead, Lieutenant Godfrey's riflemen took advantage of every "slight undulation in the surface" for cover. "The shot came through us pretty fast and quick," Lieutenant Godfrey was quoted as saying in Lord Raglan's dispatch after the battle. After a little while, they had approached to about 600 yards from the Russian artillery and opened fire with their Minié rifles.

For a few minutes, it was a contest to the death between an entire battery of Russian artillery and a handful of riflemen 600 yards away. The Russians must have been astonished to hear the rotary whip of rifle bullets about them. If the Russian guns were bronze (as most if not all the Russian guns at Balaclava were), the impact of a rifle bullet against the gun barrels would cause them to clang like a martial bell. Now certainly a little more animated, the Russians ceased their methodical, precise firing and served their field pieces with a greater sense of purpose. The range of 600 yards was too far for "grape" (canister), which was the

---

[29] Ibid.

primary type of ordnance deployed against infantry that were close enough to threaten the gunners (e.g. under 300 yards). Instead they would be firing case shot at this range, a shell filled with balls and a very small charge of powder designed to simply crack the shell (the "case") open so that the balls inside, moving at high velocity, would spread and scatter about the target. But these small Russian smoothbore guns were still using wooden fuzes with their case shot, essentially unchanged from the time of Borodino, and they were increasingly less effective as the distances grew. The effectiveness of the fuzes depended on the skill of the gunner, who cut them to the necessary length before inserting into the shell. Even if the case shot burst at 700 yards, the velocity of the ordnance was much reduced and the scattered balls and fragments often fell to earth "spent," or too slow to cause serious wounds. Case shot was still effective at longer ranges at large targets, such as maneuvering bodies of infantry, but a platoon of riflemen pressed close to earth and taking cover presented a very challenging target indeed. Beyond case shot range, artillery was forced to use the plain solid cannonball which was still effective against dense formations of troops but much less so against a small group of riflemen lying prone 600 yards away. Not only would the Russian battery be receiving effective fire from infantry with rifles, but their adversaries had placed themselves nearly beyond the effective range of the artillery's most destructive types of ammunition.[30]

The British riflemen kept up their fire. At some point in the twenty minute duel, Lieutenant Godfrey's "men handed him their rifles as they loaded them."[31] In this way, Lieutenant Godfrey could fire as many as ten shots per minute. Trained in rifle

---

[30] Bormann, Charles, *The Shrapnel Shell in England and Belgium* (Brussels, Librarie Europenne:1858), 70. Most field artillery pieces in the Russian service were light, small caliber guns intended to be hastily deployed in battle, to fire on the enemy at fairly short ranges (300-400 yards). Their effective range was rather less than contemporary British artillery. It is reasonable to assume the guns Godfrey engaged were such typical pieces.

[31] *Colburn's United Service Magazine, 1856 Part II* (London, Hurst and Blackett: 1856), 458

shooting for the entirety of his adult life, Godfrey was an exceptional shot. The Russians crouched behind their artillery pieces for cover, and whenever they would come out to try to load their guns, Godfrey picked them off, one by one. Eventually the Russian artillery ceased fire altogether, and twenty minutes after the British rifles had opened fire, the remaining Russians hauled away their guns and retired. Mere infantry had not only silenced an artillery battery, but they had driven it away from a distance of 600 yards. Nothing like it had ever happened before in the history of warfare. "We got the credit of silencing them," Lieutenant Godfrey said after the engagement. "None of our men were hurt."

This achievement was widely reported (none lesser than Lord Raglan himself gave a detailed account in his official dispatch on the Battle of Balaclava). Lieutenant Godfrey's exploit was perhaps seized upon as a comforting counterweight to what Russell colorfully described as the "glorious catastrophe," the charge of the Light Brigade. The account of Godfrey silencing the guns at Balaclava was raised up as a worthy example to follow for the Rifle Volunteer movement; England was safe from invasion if a volunteer corps of intelligent middle-class riflemen, who obtained and maintained their own arms, trained to the same pitch of excellence as Lieutenant Godfrey, stood prepared to defend her. Hans Busk, one of the prominent founders of the Rifle Volunteers, wrote "the fact should be indelibly recorded, not only in justice to Lieut Godfrey's presence of mind and gallantry, but as serving to show that a single rifleman may even at 600 yards silence artillery."[32]

Soon, however, British infantry armed with Minié rifles silencing Russian artillery became almost commonplace as the siege of Sevastopol wore on. Captain Archibald Alison (later Sir Archibald and a colonel) of the 72nd Highlanders was sent out "with a few picked marksmen" to silence heavy Russian guns that had been laying a very effective fire. Expertly judging the distance

---

[32] Busk, Hans, *The Rifleman's Manual; or, Rifles and How to Use Them* (London: 1858), 59.

as 600 yards, and instructing his men to set their sights accordingly, Captain Alison's group of riflemen "entirely silenced the battery in thirty minutes."[33] Even before Lieutenant Godfrey's exploit at Balaclava, the Minié rifle's utility against artillery had been appreciated in an order for the entire First Division issued on October 16, 1854, just as the Allied forces were establishing their siege lines. (This order was five days after Lieutenant Godfrey had distinguished himself by holding a small outpost against a determined Russian sortie, and using the Minié rifle to great effect.) Under the divisional order, ten soldiers and a sergeant from each battalion (each "good shots") were assigned to energetic junior officers with orders to "approach within 400 or 500 yards of the enemy's works." They were specifically directed to deploy "under cover of anything that may present itself to afford protection" and to remain in place for as long as 24 hours. Much like modern soldiers, "each man will select the spot that suits him best." Once in place, and behind suitable cover, they were to "endeavour to pick off the enemy's artillerymen in the embrasures." Captain Gerald Goodlake, later Lieutenant General Goodlake, VC, first distinguished himself as the leader of one of these elements that "picked off" Russian artillerymen from 500 yards. This manner of "sniping" at enemies inside their entrenchments was not particularly new (even the word "sniping" to refer to sharpshooters had been in use since the 1820s) but the longer ranges certainly were. For the Russian gunners at Sebastopol, the risk of getting "picked off" was constant, but only a minor hindrance in the larger scheme. At Balaclava, on October 25, 1854, the remarkable achievement of Lieutenant Godfrey was approaching, engaging, silencing, and driving away a battery of *field artillery* on a battlefield: this portended a fundamental change in tactics and the traditional roles of the branches.

A little over a month later, Lieutenant Godfrey was dead. In the trenches outside Sevastopol, in the most appalling conditions, fever and cholera ripped through the British regiments. The

---

[33] Dougall, 38.

affliction did not even spare young officers like Godfrey -- "he has fallen victim to another greater scourge than war," Clifford wrote following the 25-year-old lieutenant's death. In another painful, ironic twist of history, Godfrey died of cholera shortly after Dr John Snow (investigating the 1854 Broad Street epidemic) discovered that the disease was caused by pathogen-contaminated water. The British Army had adopted modern weapons, but Godfrey and thousands of others in the Crimea died from the lack of the most basic sanitation concepts. He became just another soldier of the Queen who, for one reason or another, would never again see England. "Thank God," Clifford said of Godfrey in his diary, "he was a most temperate and good Christian. No one ever heard him say a word against anyone, and he was beloved by all of us."[34] His grave was marked and well maintained in the British cemetery on Cathcart's Hill outside Sevastopol until the Russian Revolution, when the markers were destroyed by Bolsheviks purging the Crimea of imperialist relics. A forgotten monument to him, erected by his parents, can be found at Ash Church in Kent. Now he is encountered only in dusty old texts by a handful of historians researching the arcane details of a war nobody remembers. Arthur William Godfrey is just another pointlessly shortened life among millions from every side and from under every flag, mute testament to the awfulness that is war. Lest we forget.

---

[34] Clifford, 109

4
INKERMAN AND SEVASTOPOL: A NEW KIND OF BATTLE

In the trenches outside of Sevastopol, both sides employed rifles and were painfully introduced to a new military pestilence: sniping. Riflemen were used, as earlier mentioned, to "pick off" Russian gunners attempting to serve their pieces along the defenses. This fire was so effective that the gunners in the defenses were driven to form shields out of rope (an abundant commodity in a large naval base like Sevastopol) to prevent rifle bullets entering through the embrasures. The new tactical options and innovations revealed by the rifle were being quickly thought through. Rifles demonstrated "in a remarkable way the power of their weapon against the great guns," George Dodd wrote in his 1856 *Pictorial History of the Russian War.* "It was found that a rifle-pit was in effect a battery which, at 600 or 700 yards distance from a battery of large ordnance, had the power to drive away the artillerymen from the embrasures at which the guns appeared."[35]

Russell mentions Rifleman Hubert of the 2nd Battalion, Rifle Brigade, dropping "astonished" Cossacks at 900 yards on October 3, 1854.[36] A few days later, on October 15, shortly after the trenches were opened at Sevastopol, four soldiers of the Rifle Brigade "crept up to within 500 yards" of the main Russian barracks building in the city. It must have been a great shock for the Russians in the barracks when P1851 rifle bullets began crashing through the windows.[37] The Russians also employed their own rifles, with excellent effect. A British engineer officer's

---

[35] Dodd, George, *Pictorial History of the Russian War* (London: 1856), 404
[36] Russell, William Howard, *The War: From the Landing at Gallipoli to the Death of Lord Raglan* (London: 1855), 198.
[37] Verner, Willoughby, *The Rifle Brigade Chronicle for 1895* (London: 1896), 20.

letter, published in the *Morning Chronicle's* 30 April 1855 edition, said that "the greatest foe we have is the Russian rifle-bullet. It is not so perfect as ours, but, as Mercutio says: 'it will do well enough.' This little gentleman gives you no warning but flies all day long and ranges 1200 yards."[38]

A final Russian attempt to achieve victory in the field was defeated at Inkerman, on November 5, 1854. A massive Russian force of 35,000 or more troops, formed into two powerful columns, attacked the flank of the British lines at Sevastopol; possibly as few as 10,000 British troops were engaged, with a handful of French forces arriving in time to assist. Inkerman was the bloodiest battle of the war, by far, and Russian casualties were more than double those of the French and British allies. The Minié rifle is credited for much of the superior performance of the British regiments in the hopelessly confused "Soldier's Battle," so named because command and control broke down and left small groups of soldiers to fight on their own. Much of the battle was also fought in a dense fog. Even at close ranges in this chaotic battle, Russian soldiers with smoothbores (often flintlocks) were at a marked disadvantage against British rifles. Repeatedly, Russian columns appeared through the fog and were pelted with Minié bullets by heavily outnumbered British defenders.

William Howard Russell credited the Minié rifle for victory at Inkerman. Without it, the woefully outnumbered and unprepared British defenses must have been overwhelmed by the massive Napoleonic columns in determined assaults. "The Russians advanced mass after mass of infantry," Russell explained. "As fast as one column was broken and repulsed, another took its place."[39] These assaults continued for three hours, with converging columns of densely packed Russian infantry advancing simultaneously, with little room to maneuver. "There was no space for such vast numbers to deploy," Edward

---

[38] Morning Chronicle, 30 April 1855
[39] Russell, William Howard, *The War: From the Landing at Gallipoli to the Death of Lord Raglan* (London: 1855), 252.

Nolan wrote in his 1857 *Illustrated History of the War against Russia,* "and therefore the fire of British Minié muskets and artillery made havoc unparalleled in the masses exposed to such a relentless fire."[40] We were "provided with a superior rifle," the historian of the Coldstream Guards observed, and so "the enemy, emerging from the ravines, found himself met with a heavy and shattering fire."[41]

Fog and confusion defined Inkerman, but whenever the mist lifted to allow visibility of a few hundred yards, British rifle fire ripped through the Russian infantry. The Minié rifle bullet was a cylindrical elongated projectile, expanded to .702-inch diameter when fired and weighing well over an ounce; such a heavy bullet carried vastly more energy than the old round musket ball. At close range, Minié bullets could blow through multiple soldiers crammed into densely packed columns. Nolan wrote as the mass of Russians advanced into "deathful volleys of Minié musketry" there was "scarcely a shot that did not put two, and many three men *hors de combat* -- the Minié balls, piercing the bodies of the nearer soldiers, entered those in their rear."[42] The losses inflicted on the overwhelming Russian masses, by relatively small numbers of British troops with the Minié rifle, astonished Russell. "The Minié is the king of weapons -- Inkerman proved it," Russell extolled in 1855. "The regiments of the Fourth Division and the Marines, armed with the old and much-belauded Brown Bess, could do nothing against the Muscovite infantry, but the volleys of the Minié cleft them like the hand of the Destroying Angel, and they fell like leaves in autumn before them."[43]

Possessed with the advantages of numbers and fighting on home soil, the Russian strategy of attacking at Balaclava and Inkerman was sound. The tactical execution, however, was

---

[40] Nolan, Edward Henry, *Illustrated History of the War against Russia, Volume I* (London: 1857), 588.
[41] Ross-of-Bladensburg, Sir John Foster George, *The Coldstream Guards in the Crimea* (London: 1897), 190.
[42] Ibid., 569
[43] Russell, *The War*, 253.

flawed due to failures in planning and communication between the Russian elements. The Minié rifle did not, by itself, condemn the Russians to inevitable defeat, but it was a powerful contributor. Russian casualties had already been catastrophic up to this point in the war, and the well-equipped and experienced Russian soldiers the Allies encountered at Alma had, by Inkerman only a few months later, been replaced by poorly equipped and inexperienced conscripts. Period British accounts of Inkerman insist the Russians were inebriated to a man. Drunk or not, the Russian attacks at Inkerman lacked nothing for sheer audacious courage, as the columns attacked, were repulsed, and reformed to attack again over the bodies of their own dead and wounded. Attacking at Inkerman was a sound strategic decision, but it was doomed by the inexplicable incompetence of the Russian command, who could not have executed an attack more poorly. Even so, sheer weight of numbers and the nearly unstoppable force of the Napoleonic columns employed should, by conventional standards, have carried the day. The Russians themselves attributed defeat at Inkerman, at least partially, to the Minié rifle.

"Soimonoff's division, reached even on its march by the long range weapons of the enemy, had suffered severe loss" before they even began the final assault, according to a Russian pretending not so convincingly to write an account of the Battle of Inkerman under a German pseudonym. "[We] suffered so much from their Minié rifles, especially in officers and leaders, that they fell into disorder and were forced to retreat."[44] Vastly outnumbered British forces, armed with the Minié rifle, were able to inflict wildly disproportionate casualties on their Russian enemies. Over 20 Russian battalions under Soimonov, some 13,500 men, had converged upon a relatively narrow portion of the British line that was lightly defended (although British period sources tend to exaggerate just how outnumbered they were). A few hours later, Soimonov was dead and this overwhelming force

---

[44] Murray, John (trans.), *The Russian Account of the Battle of Inkerman* (London: 1856), translated from the original published in Berlin, 1855.

had been stopped.

Eduard Todleben was a Baltic German who, as a Russian engineer officer, is famous for his brilliantly improvised defenses at Sevastopol; he became one of the few Russian heroes from the war. It is remarkable that Todleben, who is probably the highest Russian authority on the matter, credits the British P1851 Minié rifle for the Russian defeat at Inkerman. "At the beginning of the battle [of Inkerman], the infantry immediately felt the effect of superior English arms," Todleben wrote in his massive, meticulously detailed two-volume *Defense of Sevastopol*. "English rifle-armed troops commenced fire at long distances, inflicting considerable carnage, before our troops could approach near enough to use our muskets. Before approaching close enough to use muskets, our troops had already suffered the loss of many officers, which of course weakened the strength and cohesiveness [of the attack]."[45] Russian artillery was also ineffective at Inkerman, finding themselves pinned down and shot to pieces from British rifle fire at long range (conversely, the British artillery was very effective at supporting the precarious defenses at Inkerman). Todleben blames the British rifleman for silencing the Russian batteries, and preventing them from supporting the Russian attack columns. "It was more the fire of rifled small arms than that of the artillery of the enemy which reached our artillerymen," Todleben wrote, "of whom the greater part were killed or wounded." He credited British riflemen with inflicting "enormous losses" upon the Russian artillery. "A perfect cloud of riflemen" maintained a steady "very violent and very accurate" fire upon the Russian guns from the distance of 800 paces ("шагов" in the Russian, an antiquated measure slightly shorter than a yard and similar to the old Prussian *Schritt*). Desperately, the Russian gunners tried to fire back and "from time to time rained case upon them," but the artillery fire would only slow down the rifle fire for a brief moment before it resumed.[46]

---

[45] Todleben, 460.
[46] Owen, Charles Henry, *The Principles and Practice of Modern Artillery* (London: 1871), 343

We have seen the rifle used decisively at Alma and Balaclava; with devastating effect at Inkerman; and extensively in the trenches at Sevastopol. British (and French) military authorities and strategists, applying the lessons of the Crimea, were thinking out the future application of the rifle on the battlefield, with its effect on cavalry and artillery. By the end of the Crimean War, however, the allied armies (and particularly the British Army) encamped at Sevastopol had changed dramatically. By and large, by the end of 1855 the regiments of thoroughly drilled professional soldiers that landed in September 1854 were gone, decimated primarily by disease and exposure. In their place, inexperienced new soldiers fresh from the depots in Britain arrived to take their place. The rifle-musket might be the first modern infantry weapon, rushed into battle in the Crimea by the forward-thinking demands of Lord Hardinge, but it had been issued to an army woefully lacking in that most fundamental aspect of modern war: logistics. After the professional core of the British Army died of mostly preventable disease or untreated wounds, Britain now "in the midst of a hot war has to train a reserve force to encounter the evidently well commanded, well disciplined, and numerous hosts of Russia."[47] In a letter from January 1855, Clifford mourned the loss of trained, professional soldiers to battle and disease. "I look with anything but satisfaction on the children and boys now sent to fill up the places of the trained soldiers fallen in battle, or carried off by disease and overwork. These drafts cannot stand the hardships of the campaign. They die like rotten sheep. They have no more idea of a Minié rifle than they have of a theolodite. Will they be able to use it?"[48]

Clifford's question -- will they be able to use it? -- is, indeed, the fundamental question asked by soldiers and commentators in the mid-19th century about the rifle-musket. Historians ask today, "*Were* they able to use it?" The prevailing answer by

---

[47] Le Couteur, 4.
[48] Clifford, 151. A *theolodite* is a precision instrument used for measuring angles in surveying.

contemporary military historians of the American Civil War is a resounding *No*. The argument is compellingly made by Earl Hess, Paddy Griffith, et al., that for all the rifle-musket's theoretical long-range capabilities, it was too difficult to use effectively at longer range. Therefore, it was used in actual combat essentially like the smoothbore muskets of the age before, firing at ranges so close the soldier could not miss. And so, most contemporary historians argue, the rifle-musket could not be the first modern infantry weapon, and it failed to revolutionize battlefield tactics. Of course, this stands in stark contrast to the effect of the rifle-musket in the Crimean War and Indian Rebellion, where in the hands of trained soldiers it *did* revolutionize battlefield tactics. The question Clifford asked hinges completely on one fundamental aspect: to be able to use the rifle-musket, *a soldier had to be trained*.

It was well understood by 19th century military authorities from every European power that the rifle-musket was a challenging weapon to use effectively. They also understood that before a soldier could utilize the rifle-musket to the extent of its capabilities, he had to be well trained. This was never in question. With a few significant exceptions, most authorities in the British Army in the late 1850s came to believe that the value of the rifle-musket was worth the additional time, expense, and effort of training soldiers to use it properly. Thackeray, who has already been quoted, put the situation plainly: "The fire of infantry being much more effective, and at greater distances, their value in the field, in comparison with cavalry and artillery, will be very materially augmented." He immediately qualifies his assertion, writing further that "This view of the matter however is dependent on the supposition that infantry render themselves as efficient in the use of the rifle as the rifle owing to its present improvement is efficient."[49]

There was never any sort of delusion, among period authorities, that untrained soldiers would turn in their smoothbore muskets and be immediately proficient with the rifle.

---

[49] Thackeray, xiv.

"Even the Enfield itself is too delicate for raw troops," Dougall wrote in his *The Rifle Simplified* in 1859, "and soldiers now must be educated up to their weapons."[50] Arthur Walker, a lieutenant of the 79th Highlanders and an officer on the staff at the Hythe School of Musketry, argued that the fire of infantry with P1853 Enfield rifles was "at least four-fold more effective than it was formerly; provided always, however, that the training of the soldier in whose hands this weapon is placed has been such as will enable him to elicit its full power."[51] A zealous advocate for the rifle-musket, even Lieutenant Walker readily observed that "without adequate instruction and training in the soldier, the rifle must be as useless as the old discarded smooth-bores of our fathers."[52] Lieutenant Henry Watson, in a transcribed lecture at Hythe published 1862, said plainly "if the Enfield rifle is placed in the hands of an unskilled man, the accuracy and power of the gun will be thrown away."[53]

Today, we do not think twice about the necessity of extensively training soldiers in the use of their weapons. Yet contemporary historians (usually considering only the context of the American Civil War) malign the rifle-musket simply because American armies of volunteers, with little or no musketry instruction, did not use the weapon to its potential. Rather than impressing British observers with the rifle's apparent lack of effectiveness in the American Civil War, it merely reinforced that which British authorities already knew to be the case! The P1853 Enfield rifle-musket (and its contemporaries) is not an intuitive "user-friendly" weapon, at least when it comes to accurate shooting at ranges beyond 200 yards. "It requires a considerable amount of skill to do the Enfield rifle justice, even at known

---

[50] Dougall, 34.
[51] Walker, 5. The claim that the rifle was "four-fold" more effective than the old musket was often repeated in other sources, including by the American Lt. Cadmus Wilcox.
[52] Ibid., 22.
[53] Watson, Henry, *Eight Lectures Delivered at the School of Musketry, Hythe* (London: 1862), 7.

distances" Lieutenant Watson said in one of his 1862 lectures. "On service, when judging your own distance enters the consideration, we see at once that it requires a great amount of skill to make the most of a rifle."[54] To have any hope of hitting targets at any distance, the soldier had to be instructed on the use of the sights, adjusting the sights for targets at various distances, and the fundamentals of proper rifle-shooting. Without this specialized training, the spiraling grooves in the barrel provided no great advantage over the smoothbore. Is this fact alone, that inexperienced recruits without training were unable to use a modern weapon effectively, sufficient to declare the rifle-musket ineffective as a weapon type? By and large, that is the verdict of most modern historians in their treatment of the rifle-musket in the American Civil War. To some extent, modern historians concede that the rifle-musket was not used effectively because the soldiers in the War Between the States were virtually untrained. Overall, however, the reader of popular contemporary works on the subject (such as Dr. Earl Hess's recent book *The Rifle Musket in Civil War Combat: Reality and Myth*) cannot escape their conclusion that the rifle-musket itself was inherently and intrinsically too difficult to use, and its trajectory too parabolic, to be an effective weapon type beyond the point-blank range. Their criticism of the rifle-musket is so scathing, that one is left wondering why the armies of the mid-19th century even bothered replacing the effective, wonderful smoothbore musket with the difficult, finicky, and inefficient rifle.

    The transition from smoothbore musket to rifle-musket is among the greatest technological leaps in military history. This transition has received very little scholarly attention, but deserves a much closer study. In the 19th century, new soldiers could be issued a smoothbore musket and be taught how to load and fire the weapon in a few minutes; the rest of the recruit's instruction would be in *drill* and movement in formations. Hardly a thought would be given to any further instruction (let alone target practice) in the use of the smoothbore musket. As Lieutenant

---

[54] Watson, 5.

Watson said, "A few years ago our infantry were armed with smoothbore muskets. The smooth bore had very little accuracy of shooting, and very little skill was required to make the most of it; in fact, so little, that it did not appear worthwhile to have any detailed system of musketry instruction. But now the case is

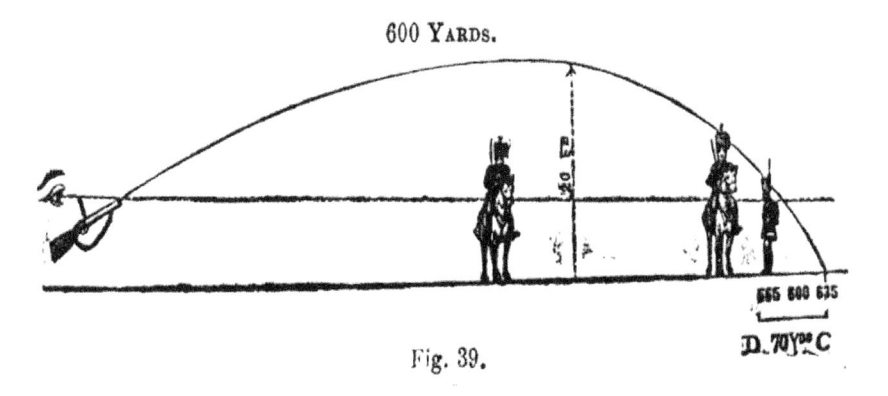

600 YARDS.

Fig. 39.

widely different."[55] With the technological leap from musket to rifle, *the soldier also had to accompany the weapon on that leap*. The same new soldier, no matter how enthusiastic, would need at least a reasonable period of comprehensive instruction before he could exploit the long-range capabilities of the rifle-musket. This specific instruction was necessary for two reasons: the pronounced parabolic trajectory of the rifle-musket bullet, and the comparatively lower muzzle velocity of the bullet.

The trajectory of a rifle-musket bullet is well illustrated by period sketches used for musketry instruction. A bullet fired from the P1853 Enfield rifle, at a target 600 yards away, reaches a height of some 20 feet above the surface at the apex of the trajectory. Then the bullet begins a fall back to earth. Assuming the soldier had perfect aim, and perfectly estimated the distance to his target as 600 yards, and set the sights on the rifle to 600 yards, as the bullet came down it would graze the top of the enemy's head at 565 yards, and hit the enemy's toes at 625 yards. This created a "dangerous space" of only 70 yards in which an enemy would be hit. To hit an enemy soldier "center mass" at

---

[55] Ibid., 6.

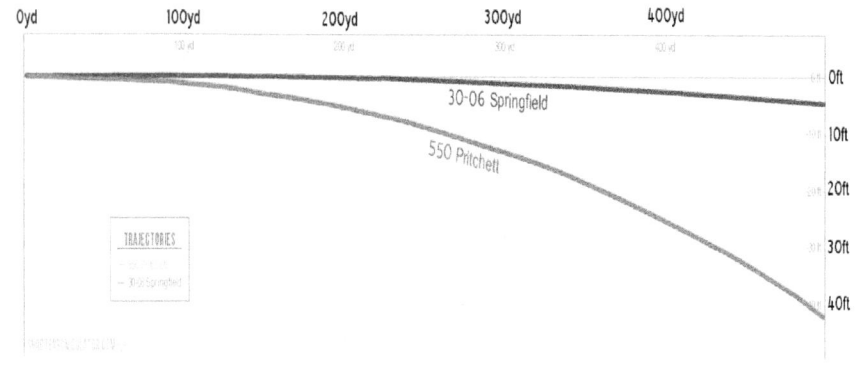

```
0yd        100yd        200yd        300yd        400yd
                                                                    0ft
                              30-06 Springfield
                                                                    10ft
                                   550 Pritchett
                                                                    20ft

        TRAJECTORIES                                                30ft

                                                                    40ft
```

600 yards required not just precise aim, but a correct estimation of the range to within 20 or 30 yards.

The pronounced trajectory arc of a rifle-musket bullet is particularly evident when compared to a more modern bullet. In the case of the graph, the .550-caliber "Pritchett" bullet of the P1853 Enfield rifle-musket is compared to the 30-06 cartridge, used by the American M1903 Springfield rifle and, later, the M1 Garand rifle. In 500 yards, the 30-06 bullet drops a total of approximately 5 feet while the rifle-musket bullet drops *over 40*. The rifle-musket's comparatively poor performance is due to the use of black powder for propellant, as modern smokeless powders (capable of doubling or even tripling the bullet's speed) would not be invented for another 20 years.

Rifle-musket bullets, being elongated cylinders, were also heavier than the round ball used in smoothbore muskets. This combination of a heavier piece of metal pushed by a relatively weak propellant meant that the muzzle velocity of a rifle-musket bullet was noticeably slower than a smoothbore musket.[56]

At closer ranges, the soldier did not need to bother with range estimation or take into account the great arc of the bullet trajectory. With sights set at 100 yards -- the lowest possible range setting on the British and American standard issue rifle-muskets -- the bullet would fly a total distance of 195 yards before hitting the ground, or the "first graze" as period sources described it. Enemy troops, from the muzzle of the gun to a distance 195

---

[56] Muzzle velocity is the speed at which the bullet is moving when it leaves the barrel of a gun.

yards away, were within the "dangerous space." In other words, from 0 to 195 yards is the distance at which even the most inexperienced soldier, without any musketry instruction whatsoever and without even touching the rifle's sights, could still hit the enemy. To hit an enemy soldier at 200 yards would require the shooter to estimate the range, and then adjustment the rifle sights to the 200 yard increment. The sight adjustment would raise the muzzle of the rifle, adding elevation to the bullet's

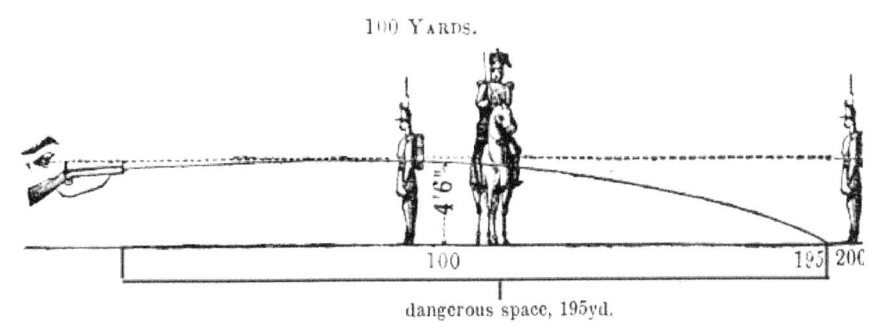

100 YARDS.

dangerous space, 195yd.

trajectory so that it could cover more distance before gravity and atmospheric resistance brought it back down to earth. It is worth noting that 200 yards is also about the maximum effective range of a smoothbore musket. The inexperienced, untrained soldier with either a rifle-musket or smoothbore musket could put effective fire on the enemy up to about 200 yards away; beyond that range, with the rifle-musket, the soldier some degree of additional training to be effective.

Mid-19th century military authorities had a profound decision to make. Were the new theoretical capabilities of the rifle-musket enough to justify an expensive program of marksmanship instruction, which required skilled instructors, vast quantities of ammunition, and considerable time? The European armies of the period were largely drawn from the illiterate peasant class. Were these uneducated common soldiers even capable of comprehending the complicated marksmanship concepts of trajectory, line of sight, judging distance, and shooting fundamentals? Most European powers did not think so. The Minié rifle was remarkable, but its powers were beyond the grasp of the ordinary conscripted peasant soldier. Only two European militaries made the difficult, controversial, and expensive decision to commit to the *rifle* as the infantry weapon of the future: Great Britain, and the small Kingdom of Prussia.

5
FIRE TACTICS: A PARADIGM SHIFT

The sole advantage the rifle-musket provided to the armies around the year 1850 was the *potential* for trained, proficient soldiers to hit targets at ranges beyond the reach of old smoothbore musket. In all other respects, the rifle-musket was nothing new. It weighed generally the same, it utilized the same bayonet, and it had the same rate of fire (three rounds in a minute was the very best that could be expected in battle from a remarkably well-trained an extremely steady soldier). Yet the rifle-musket gave infantry the ability to effectively engage the enemy at longer distances, and this opened up a new possibility in European warfare: it was now possible to defeat the enemy with gunfire alone from a distance, where previously it almost inevitably required a combination of the bullet and bayonet at close quarters.

We take it for granted today, as inhabitants of the 21st century, that infantry soldiers employ *firepower* from rifles and machine guns as the means of engaging the enemy on the battlefield. It is easy to forget that the question was by no means settled in the mid-19th century. Movies, popular history, and a uniquely American gun culture has built the impression that as soon as infantry were armed with firearms, they fought battles by shooting projectiles back and forth at each other. In this popular view, close hand to hand combat was abolished, except for a few rare instances. Some of these impressions are older than the rifle-musket itself: Colonel Wilford often spoke of the American sharpshooters that defeated the disciplined British forces with long-range fire in the War of Independence. The enduring myth that the Americans won the Revolutionary War because hardy backwoodsmen with Kentucky rifles hid behind trees and boulders and picked off the British from long range took hold early and never let go. But the fact remains that even into the latter years of the 1860s, the infantry's ability to fire bullets and engage opponents at a distance was not universally seen as the *primary* means of infantry combat. The bayonet was very much still esteemed as *the* most effective, decisive, and fundamental weapon of the infantry in the doctrine of many European armies, particularly in France, where it was enshrined in the victories of Napoleon and spelled out in the dictums of Jomini. Musketry fire could shape or influence the battle, but the final push with the bayonet was paramount.

The German word *Feuertaktik* or "fire tactics" describes the mode of combat in which the infantry engages enemies at a distance, using the firepower of muskets or rifles. In fact, the term *Feuertaktik* predates the rifle-musket era, and evolved over the second half of the 19th century to describe infantry projectile combat fought solely with firearms, instead of simply using musketry as a preparatory tactic prior to a bayonet charge. Infantry using *Feuertaktik* will fire at the enemy with effect, at a distance, with the intent of destroying the enemy (or causing them to break) without resorting to the bayonet or any other means. There is no English-language word equivalent to *Feuertaktik*, and the closest approximation of "fire tactics" fails to convey the fuller meaning of the German. "Decisively engaging and defeating the enemy with musketry alone" is a more useful dynamic equivalent translation. Today, the first historical association of the term *Feuertaktik* is with General Helmuth von Moltke (the Elder) and the Prussian war with Austria in 1866, where the high volume of fire from the bolt-action Dreyse breechloading needle-rifle contributed substantially to the Prussian victory. The Prussians completely smashed the Austrians at Königgrätz almost exclusively with infantry fire, and scarcely a Prussian bayonet needed to be fixed. Moltke, Prussian *Feuertaktik,* the needle-rifle, and Königgrätz have been distilled together by popular military history as *the* transition from the Napoleonic era to the modern. While Moltke's Prussians get the historical recognition, the British Army, with the Pattern 1853 rifle-musket, had adopted a form of *Feuertaktik* almost a decade before the needle-rifle demonstrated the superiority of breechloading firepower at Königgrätz.

In many ways, the British Army of the 1850s lived up to the stereotypes. Officers frequently purchased their commissions, and except for officers going into very specialized or technical branches, there was no examination process. The military organization was inefficient and hidebound by the decades-long retention of senior leaders who, like the Napiers and Wellington, were brilliant and successful in their day, but were hesitant to change the system that had worked so well in 1815. They were skeptical of the advance of technology that challenged the "universal" military principles of Napoleonic perfection that Jomini had turned into a sacred writ; it was considered much safer to continue to trust in the great principles of war that had been demonstrated successfully during the Napoleonic wars, rather than explore new and untested technology, principles, and theories. A woefully inefficient and complicated system of administration had a Kafkaesque number of offices and overlapping authorities, and the abysmal Army administration during the Crimean War inspired the "Circumlocution Office" satirized in Dickens's *Little Dorrit*. Yet out of this wretched morass, almost in spite of itself, the British Army recognized the capabilities of modern rifled weapons, adopted them universally across the entire rank and file of the infantry, and sought to make every infantry soldier proficient in their use. The British did this *years*, and in some cases *decades*, before any other nation. There was robust debate and inveterate resistance to this transition, but during the period of 1853 to 1860, the British Army as an institution embraced the rifle-musket, and decided that future battles and wars would be decided by firepower: not the bayonet, not the charge, not shock, but *bullets* fired by trained soldiers would be tactically decisive on the future battlefield. This subtle but significant transition announced a monumental paradigm shift in land battle and the role, status, and ability of the common soldier.

Until the adoption of the rifle-musket as the general-issue weapon for all ranks, it was almost impossible to actually defeat an enemy army using musketry alone. The short range of the wildly inaccurate smoothbore musket meant that the enemy would already be quite close before firing could begin with any effect. A rapidly advancing attacking force would, therefore, receive at most one or possibly two volleys of musketry before bayonets crossed. Napoleon Bonaparte raised large conscript armies and routed the standing armies of conventional line infantry by attacking in rapidly-moving column formations. A few blasts of musketry simply could not stop a dense column of assaulting infantry that was propelled forward inexorably by its own mass, delivering overwhelming numbers of troops at the decisive moment and place, each with a loaded musket and (more importantly) a bayonet. These Napoleonic shock tactics were enormously successful, and they adapted well to large, conscripted armies. The individual training and proficiency of the common soldier were subordinate to the principles and strategic geometry codified by Jomini, which held that the general who decisively concentrates his troops in time and space would win.

The British Army had two unique aspects that likely predisposed the army's adoption of the rifle. It was a volunteer army of professional soldiers who performed long periods of service, instead of the short-term conscripts of most continental powers. Just as importantly, the 19th century British Army was enormously proud of its reputation as a "solid" army that stood in line formations and unleashed a crushingly effective blast of smoothbore musketry that stopped countless French assault columns during the Napoleonic Wars. It might be successfully argued that a primitive form of *Feuertaktik* was already present in the British military culture long before the advent of the Minié rifle. Solid British lines, formed in just two ranks, delivered crushing musketry fire from Brown Bess that gnawed into the attacking French columns. The British, often outnumbered, would whittle down the odds with these blasts of musketry, and then bayonet charge into the stalled, bloodied, shaky French formation, which would obligingly break and rout.

Thomas Robert Bugeaud, a Marshal of France who would

achieve his fame fighting in Africa in the 1840s, commanded a French battalion during the Peninsular campaign. As quoted in Oman's excellent essay *Column and Line in the Peninsular War*, Bugeaud described the stoic resolve of the British line, and the physical and moral effect of British musketry fire upon a French column of attack:

> The English, quite silent, with grounded arms, looked, in their impassive immobility, much like a long brick wall; their aspect was imposing, and never failed to impress young soldiers. Soon as the distances began to get shorter frequent cries of *"Vive L'Empereur, en avant, à la baïonnette"* began to be heard: some men hoisted their shakos on their muskets, the quick-step became a run, the ranks tended to melt into each other, the agitation became tumultuous, many soldiers began to fire as they ran. The English line, still silent and unmoved, with grounded arms even when we were only three hundred yards away, seemed to take no notice of the storm which was about to beat upon it. The contrast was very striking. Many of us began to reflect that the enemy was very slow at starting his fire, and calculated that the fire so long held back would be very unpleasant when it did commence. At this moment of painful expectation the English line made a quarter turn -- the muskets were going up to the "ready." An indefinable sensation stopped many of our men dead: they halted and began a wavering fire. The enemy's return, delivered with simultaneous precision, absolutely blasted us. Decimated by it, we reeled together trying to recover our equilibrium. Then three formidable hurrahs! terminated the long silence of our adversaries. At the third they were upon us pressing us into a disorderly retreat.[57]

---

[57] Oman, Charles W. C., *Column and Line in the Peninsular War* (London: Oxford University Press), 21

This scene was repeated time and time again during the Peninsular campaign, and finally again on the grandest possible scale at Waterloo. Sir William Patrick Francis Napier's *History of the War in the Peninsula* (which was popular reading, especially among the British officer class) praised the firepower of British smoothbore musketry, concluding "the English fire is the most destructive known."[58] He was probably quoting Wellington, who had written in 1835, "I considered our Arm [Brown Bess] as the most efficient that had yet been produced. The fire from it undoubtedly is acknowledged to be the most Destructive known."[59]

One of the clearest examples of this destructive power of British smoothbore musketry was at the Battle of Talavera in 1809, where General Arthur Wellesley was created the Viscount Wellington following his victory on that battlefield. At one central section of the British position, a brigade of infantry under Richard Stewart "deployed his men two deep in a line nine hundred yards long," which allowed all 1,500 of his brigade's muskets to be brought to bear. General François Ruffin, a French officer present at several of Napoleon's victories including Austerlitz in 1805, advanced his three veteran regiments of infantry (some 4,000 men) in battalion columns against the thin British line. These tactical French attack columns had secured victory on countless battlefields across Europe, and in formations nine ranks deep they marched up the Medellin ridge toward the British. They closed within a hundred yards before the British fired. Hit by the scythe-like volley of musket balls, the column stopped for a moment but recovered from the shock and resumed the advance. This had been just enough time for Stewart's thin red line to reload, and they fired a second volley into the mass. Again the French stopped, recovered, and tried to

---

[58] Napier, William Patrick Francis, *History of the War in the Peninsula and in the South of France* (Oxford: David Christy, 1836), 122

[59] Maxwell, Sir Herbert, *The Life of Wellington* (London: Sampson, Low, Maston and Company, 1899), 136

press forward, only to receive a third volley. Then a fourth, and a fifth. The musketry never stopped, and the French were cut down faster than they could deploy. In just three minutes, the British had fired ten volleys; even generously accounting for the misfires of flintlock muskets, anywhere from 10,000 to 12,000 musket balls had been fired into the French columns at the range of 50 to 100 yards. Another British brigade under Tilson wheeled into position and began firing into the flanks of the French, who had finally had enough. They began to fall back, and sensing the decisive moment, the British charged ahead with the bayonet into the shattered remnants of the French formations.[60]

Wellington would defeat these French columns in a string of battles on the Peninsula, and finally at Waterloo, which was the last word on large European battles for the next half-century. A pattern readily emerges when examining the battles the British fought against the French in the Napoleonic Wars: the British defend a position, the French attack it with columns that the British weaken with a coolly-delivered blast of musketry, and then the British launch a spirited bayonet assault that routs the shaken French. In the decades after 1815, the heroes of Waterloo and the Peninsula were revered, as were their methods and weapons. When the percussion lock appeared, there was general agreement that it should replace the flintlock on Brown Bess, but *all other* aspects of the weapon remained the same. Although it was worshipped as the musket that had beaten Napoleon on a hundred battlefields, some officers acknowledged that (by 1850) "with the bayonet fixed it was the shortest gun used by any European army, at the same time the heaviest; it had the most windage; fired the largest charge of powder; it had the greatest recoil; the least accuracy."[61] In addition, it could be added that Brown Bess was also the most expensive, compared to lighter, smaller-caliber muskets of contemporary powers, which utilized economical barrel bands (instead of pins and wedges) to hold the

---

[60] Parkinson, Roger, *The Peninsular War* (Hertfordshire, UK: Wordsworth, 2000), 94
[61] Wilford, 2

barrel to the stock. But this was the musket that delivered the most destructive fire known, when carried by the solid, unmovable British soldier. The British system of battle – crushing musketry to shatter the enemy, aggressive bayonet charge to finish him – surrounded Brown Bess with a legendary aura.

This was something far less than true *Feuertaktik* but it was fertile soil for advocates of the rifle. In British military culture of the first half of the 19th century, there already existed a firmly-seated conviction that the firepower of musketry was at least equal to the bayonet as the primary mode of infantry combat, and destructive musketry was essential to victory. The adulation of Brown Bess and the crushing musketry of the British infantry of the line approached a near-mythical apotheosis. The appearance of the rifle-musket, however, challenged the existing sacred paradigm. Unsurprisingly, the fountainhead of this reverential devotion was the Duke of Wellington himself, who "retained an almost superstitious admiration for Brown Bess."[62] The destructive firepower of Brown Bess, with volleys delivered every fifteen seconds, extended only to a range of perhaps 100 yards and the greatest execution was done at 50 yards or less. Yet no matter how quickly Brown Bess was fired, its destruction was limited to a relatively short span of ground. The enemy could be badly mauled by this short-range musketry, but the ultimate victory still relied on the bayonet charge at the decisive moment. The rifle-musket, on the other hand, extended the potential effective range of the infantry's firepower from 100 yards to 500 yards and beyond. With the rifle-musket, it became theoretically possible to defeat the enemy with bullets alone, with pure firepower, before the enemy got close. Instead of waiting for the enemy to close within the last 100 yards before delivering the crushing, destructive fire the British infantry was known for, early advocates of the rifle-musket imagined the same crushing, destructive fire pelting the advancing enemy columns with bullets at 500 yards or even more. The bayonet would become superfluous; the enemy would be destroyed by pure firepower at

---

[62] Maxwell, 136.

a distance, as if brushed away by the invisible hand of a destroying angel.

In theory, it was not a very audacious proposal: if British musketry was so effective at 50 yards, with the new weapons the British musketry could be just as effective at 500. Those same qualities of British pluck, solidity, discipline, and efficient fire of musketry that had defeated French columns at short range would naturally be enhanced by the capabilities of the new rifles. But the new theory required a shift in the paradigm in two essential areas. First, it required an infantryman well-trained enough to fire a weapon of precision at a target at long distance with the same ease that he could fire Brown Bess at a target 50 yards away. And second, it required a change to *how* battle was fought, and this was a gross heresy at a time when the influence of the Iron Duke and his Peninsular officers still remained extremely strong in the institutional British Army. A number of the very highest authorities dogmatically insisted that the rifle would never prove superior to the smoothbore musket in the line of battle. Behind their confident assertions about the rifle, they had no idea what a battle among rifle-armed troops would look like, or how to fight one, and they feared the unknown and the untested.

General William Napier was never shy to defend the destructive power of British musketry with Brown Bess, but he was a powerful and leading voice against entertaining any sort of *Feuertaktik* (or any other variation from Jomini's great principles) in the British service. An admirer of Napoleon, William Napier was among the first in Britain to discover Jomini, writing a review of *The Art of War* in 1821. He described Napoleon's campaigns in Italy as "the period in which the principles of the military art were brought to all the perfection of which they appear to be capable."[63] Convinced that "perfection" had already been reached, Napier was an unsurprising opponent of any proposed changes to Napoleon's rules of war. In several letters and essays, he brought his great authority to bear against adopting the rifle

---

[63] *The Edinburgh Review, or Critical Journal: for March 1821-July 1821*, Vol XXXV, 380

as the universal infantry weapon, and particularly against the new ideas of defeating enemy forces solely with firepower. The new Minié rifle, the old soldier lamented in 1853, "is likely to supersede the old musket. I am sorry for it; not sorry that the better weapon should be adopted, but sorry that the improved weapon should have been invented." William Napier condemned the newfangled weapon precisely because he sensed the paradigm shift it heralded; the rifle would "paralyse" the traditional role of cavalry and artillery, "the generals would inevitably be killed, and hence battles would be more confusedly arranged, more bloody, and less decisive." Better altogether, he argued, to simply not adopt such a weapon, and stick with Brown Bess. "It will not serve England," he insisted. Instead, "our only real superiority lies in our resolute courage to close with the bayonet" and "the fear of it in English hands which gave us victory, and the Minié rifle will probably deprive us of that moral force." In 1855, after the paradigm-shifting power of the rifle had been demonstrated on the Crimean War battlefields, William Napier was still unmoved. "One swallow does not make a summer," he insisted in a letter from January 1855, throwing the full weight of his authority behind the conclusion that "the Minié rifle is not adapted for general warfare."[64]

William Napier's brother Charles was also a British general, and had an equally impressive military career. As the commander of a British regiment of infantry during the Peninsular campaign, Charles Napier was severely wounded and left for dead; he would survive and serve extensively in high posts around the British Empire for the next 40 years, often in crippling pain. He died a lieutenant general in the Order of the Bath in 1853 and today his statue occupies one of the famed plinths at Trafalgar Square. Shortly before he died, he joined his younger brother William in chorus against the replacement of Brown Bess and the adoption of a rifle-musket. "Rifles should be abolished," he wrote around 1852, "the weapon is not good for war." Charles was in complete

---

[64] Bruce, H. A., Ed., *Life of General Sir William Napier, K.C.B.* (London: John Murray, 1864), 378

agreement with his brother about the corrosive effects of arms of precision. "As to the Minié rifles, they will in my opinion destroy that intrepid spirit that makes the British soldier always dash at his enemy." To Charles Napier, the extreme inaccuracy of Brown Bess was a marked advantage in British hands. "The short range and very uncertain flight of shot from the musket begets the necessity of closing with the enemy," he wrote,

> which the British soldier's confidence in superior bodily strength, due to climate, pushes him to do; he takes his stand in line of battle, thinking the *ne plus ultra* of glory is a close volley and a charge of bayonets, with his terrific shout of battle. Stop that and he is a common man. Frederick the Great encouraged his troops to charge first and fire after the bayonet had done its work; modern generals are seeking by the Minié rifle to do away with the employment of the bayonet.[65]

By "modern generals," Napier probably refers to General Hardinge, who in 1852 was Master-General of the Ordnance and by September of that year Commander-in-Chief of the British Army following the death of Wellington. Hardinge, as we have seen, was an advocate of the rifle and instrumental in the development of the Pattern 1851 rifle that would be the very Destroying Angel of the Crimea. At this extraordinarily early date, Charles Napier savagely kicked back against the proposed *Feuertaktik*. "The Minié rifle, perfected, will ring the [death] knell of British superiority," he prophesied. "When were French soldiers ever beaten by fire only?"[66]

Another authority of the highest reputation piled on against the rifle and the concept of fire tactics gaining ground in the British Army. Sir Howard Douglas was the son of a Royal Navy

---

[65] Napier, Sir Charles James, *Defects, Civil and Military, of the Indian Government* (London: Charles Westerton, 1857), 305
[66] Ibid., 306

admiral, a veteran of the Peninsula, and by 1851 a full general. He is best known as the writer of *A Treatise on Naval Gunnery*, first published in 1820 but periodically updated, revised, and expanded into the 1860s. The third edition, published in 1851, contained an appendix about the Minié bullet and the rifle-musket, and Douglas cautiously predicted that "this invention may prove highly useful for rifle corps and special purposes." [67] Here his potential appreciation for the rifle ends, and for several pages Douglas proceeded with a litany of arguments against the rifle-musket for line infantry use. "To abandon to any extent those simple, trustworthy, and, in war, well-tried weapons – the regulation [smoothbore] musket and ball – in favour of a complicated, and in real service untried substitute, would be a very rash experiment to be made with the infantry of the British Army." Douglas observed that the French, who had timidly adopted the rifle-musket for special troops, were still engaged in comprehensive experiments and undecided on any larger role. "Nor does it appear that other military services are in any haste to adopt this novelty," Douglas wrote.

Like many senior officers of his generation, Douglas was convinced that expensive and "delicate" weapons like rifles would be wasted on the common line infantryman. In the ranks of the infantry there were, Douglas argued, a "large proportion" of soldiers whose "constitutional nervousness and want of intelligence" disqualified them in the use of any weapon more complex than the smoothbore musket. Ordinary soldiers, he argued, "can never, with all the practice and instruction that can be given them, be made to understand the use of sights, or taught to fire correctly with the rifle, though they may do their work, when in line, as well as the best and steadiest shooter." Because the use of sights requires a basic knowledge of distance estimation, Douglas was convinced this ability was beyond the capacity of the average soldier. Judging distance "is far too

---

[67] Douglas, Sir Howard, *A Treatise on Naval Gunnery, Third Edition, Revised* (London: John Murray, 1851), 600

complicated a matter to be left to the judgement and intelligence of soldiers of the line, and is not required in line-firing." He concludes decisively: "Rifles will never answer for line-firing."[68]

The fourth edition of *Naval Gunnery* appeared in 1855. By the printing of the fourth edition, dramatic events had shaken the foundations of Douglas's assertions about rifles, and the capability of common infantrymen to use them. Lord Hardinge had succeeded in getting the Pattern 1851 Minié rifle adopted and, more importantly, issued just in time for British troops heading to the Crimea. Accounts of the Alma, Inkerman, Balaclava, the Thin Red Line, Lieutenant Godfrey, and the Siege of Sevastopol filled British newspapers; William Howard Russell's prose, praising the Minié as the King of Weapons and the Destroying Angel, was popular reading. In the fourth edition of *Naval Gunnery*, Douglas completely updated the appendix on rifle-muskets. The dogmatic exclamations about rifles *never* answering for line fire, or soldiers who could *never* be taught to use rifle sights, were gone. Instead, Douglas re-wrote the appendix with a history of the rifle-musket and condensed his objections to the adoption of a rifle as the general weapon for the infantry into a single sentence. After quoting a French authority on the destructive efficiency of British smoothbore musketry in the Napoleonic wars, Douglas "we should be very careful not to compromise that established efficiency by any general or extensive adoption of the new arms."[69] A final edition of *Naval Gunnery* was published in 1860, by which time the matter had been settled. Instead of just an appendix, Part VI of the treatise was devoted to the rifle-musket, filled with diagrams and facts and tables "kindly communicated by Colonel Hay" from the School of Musketry. If Douglas still believed the rifle would never answer for line firing, or that soldiers were not intelligent enough to use them, he kept these thoughts to himself. He died in 1861.

Between the third and fifth editions of *Naval Gunnery*, the

---

[68] Ibid., 604
[69] Douglas, Sir Howard, *A Treatise on Naval Gunnery, Fourth Edition, Revised* (London: John Murray, 1855), 513

British Army institutionally adopted fire tactics built around the increased destructive powers of the rifle-musket. This was not a clean, immediate, or even well defined process. The influence of the School of Musketry on this process was enormous, and grew exponentially over the decade of the 1850s. Colonel Charles Crawford Hay, of the Green Howards (19th Regiment of Foot), was an accomplished rifle shot; Lord Hardinge could not have selected a better officer to be of the new School of Musketry. As a determined advocate of *Feuertaktik* and arming the whole British infantry with a rifle, Colonel (later Major-General) Hay left an indelible mark on British arms, tactics, and doctrine. Hans Busk, a leader of the early Rifle Volunteer movement, referred to Hay as "that prince of marksmen" and quoted a Scottish musketry instructor, who praised Hay as "the greatest of all airthly [sic] rifle shots." Captain Augustus Henry Lane-Fox was briefly the Chief Instructor at Hythe, but he was dispatched to Malta as musketry instructor and was instrumental in training soldiers on their way to the Crimea. In his absence, Lt. Colonel Earnest C. Wilford (also from the Green Howards) was appointed Chief Instructor. Wilford, like Hay, was very energetic and a riveting, engaging lecturer. Even critics who strongly disagreed with Wilford were quick to recognize his zeal. Probably more than anyone else, Wilford's engaging, amusing, and animated teaching style left a personal and deep impression on his students. Colonel Wilford was absolutely convinced of the superiority of *Feuertaktik* and the capabilities of the rifle-musket. The officers and NCOs sent to Hythe were almost universally persuaded by Wilford's compelling lectures, and returned to their units to spread the gospel of fire tactics.

Quite simply, students at Hythe were fed an unrelenting diet of fire tactics, starting on the first day of the course. Colonel Wilford gave most of the introductory lectures himself, and General Hay made frequent appearances also. A student at Hythe in 1860 later recalled the impression made by Colonel Wilford:

> The first day was devoted to platoon and introductory

> lectures from Colonel Wilford, the principal officer of instruction. Colonel Wilford is a very soldier-like man, something over 60 years of age, tall, erect, active, and of most lively and pleasing manners. His style of lecturing is admirable; enthusiastic in his subject, he has the power of exciting enthusiasm in his hearers. A fine manly tone pervades all he says. He speaks like a Christian, not less than a soldier. He is graphic, vigorous, full of humour, abundant in illustration, and often rises to eloquence. Before him is a table spread with models behind him a black board for diagrams; and on the right is his rifle, which, when the subject invites him, he seizes, and with startling effect, suits the action to the word. I cannot hope to give you even the ghost of an idea of the vivid impression produced by Colonel Wilford's lectures.[70]

Many of Colonel Wilford's lectures were preserved in the notes of his students at Hythe, and Wilford also published a short volume of three of his lectures on the rifle. In addition to his prolific lecturing and writing, Wilford read papers at the United Services Institute, where he ruffled more than a few feathers by attacking the officers (sometimes by name) who still doubted the wisdom of replacing the infantry smoothbore with the rifle. For the better part of a decade, Colonel Wilford lectured class after class of officers and NCOs at the School of Musketry, converting the majority of them with his convincing near-religious fervor and enthusiasm for fire tactics. Infected with the gospel of Wilford and Hay, which renounced the bayonet and elevated the fire of trained riflemen, thousands of junior officers went back to their regiments with firm conviction in the supremacy of *Feuertaktik* as well as the practicality of training common soldiers to become effective marksmen. In this way, the influence of the School of Musketry was felt across the breadth of the British Army, and the preeminence of fire over the bayonet transcended

---
[70] Edwards, 6

the formal doctrine and manuals of regulation.

"By all means have a bayonet," Colonel Wilford conceded in a lecture at the School of Musketry in November 1859, "and learn how to use it to the utmost advantage; but remember, that it has long been the object of infantry to attack with their projectile."[71] He further urged his students to "remember that the bayonet only took the lead as an infantry weapon, because the gun could not be depended upon to hit." Now that the British Army had a rifle that the ordinary line soldier could use with effect at targets hundreds of yards away, the bullet rather than the bayonet must become the primary weapon of the infantry. In 1860, Colonel Wilford told students at Hythe plainly: "Infantry fight with projectiles."[72]

The 1859 revision of the British Army's primary manual of infantry tactics, *Field Exercise and Evolutions of Infantry*, officially implemented fire tactics across the entire infantry, at least in theory. In the introductory paragraphs to the new chapter on Musketry Instruction, the regulation specified that "The rifle is placed in the soldier's hands for the destruction of his enemy." The new chapter appeared at the end of the manual, following hundreds of pages on specific drill movements in lines, wheels, and columns. "In fact," the regulation specified, "all his other instructions in marching and maneuvering can do no more than place him in the best possible situation for using his weapon with effect. A soldier who cannot shoot is useless, and an encumbrance to the battalion."[73]

Lieutenant Watson, also an instructor at Hythe and a disciple of Wilford's fire tactics, told his students in 1862, "just as the object of all marching and maneuvering is to place the soldier in the best possible position to use his rifle, so the object of musketry instruction is to teach him to use his rifle with the greatest possible effect when in position." Watson emphasized that even a good soldier – with outstanding qualities of discipline,

---

[71] Wilford, 70
[72] Edwards, 10
[73] *Field Exercise and Evolutions of Infantry* (London: Horse Guards, 1859), 381

excellence at drill, clean habits, and soldierly deportment – would be of no use if he could not shoot. "If an infantry soldier is not a good shot, he is only good in a measure," Watson continued. "Our so-called 'good soldier,' if he cannot shoot, will fail when he is most wanted, and, so far from proving a good soldier, will turn out, on the day of battle, a useless encumbrance."[74]

---

[74] Watson, 6

6
THE RIFLE AND THE TRAINED SOLDIER

The greater difficulty of using a rifle-musket effectively at ranges 200 yards and beyond, as opposed to simply leveling the old smoothbore at point-blank range, adds to the argument that the rifle-musket was, indeed, the first modern infantry weapon. Consider the different expectations of the soldier where it concerns the operation of his weapon on either side of the great technological leap from smoothbore to rifle. Colonel Wilford spoke at the United Service Institution in July, 1857 (two years after the Crimean War, and four years before the outbreak of the American Civil War). In his lecture, Wilford praised the smoothbore Brown Bess as a formidable weapon at short ranges, and described the soldier using it. "In firing," Wilford said, "he shut his eyes, opened his mouth, threw his head back, and pulled the trigger; and, if this was not enough, he was sometimes exhorted to 'Aim low'."[75] The engineer and famed gunmaker General John Jacob of the Bombay Army agreed, calling British troops armed with the smoothbore "pipe-clayed automatons."[76] In stark contrast, the soldier with the rifle-musket would be a "skillful workman," according to General Jacob, harnessing the "high moral and intellectual powers" of the English rifleman. Wilford emphasized "the astonishing powers of the rifle in the hands of a taught soldier" in a succinct passage:

> A man cannot be made to hit a mark against his will, in fact, he fires with his brains, the eye and finger being merely servants of the mind. To succeed or to excel, there must be love in the heart and knowledge in the head, but no man can be interested in that which he cannot understand. Hence each soldier is made to

---

[75] Wilford, Ernest Christian, *Three Lectures upon the Rifle* (London: 1859), 11.
[76] Jacob, John, *Rifle Practice* (London: 1857), 29

comprehend the laws which influence the bullet in its flight, and how to apply this knowledge to practice. He is led to think and his moral character is found to be improved and elevated thereby. He becomes conscious of his increased efficiency and value; he is raised from a mere machine -- a trigger-puller -- a thrower away of fire -- and after instruction is not merely a good but an intelligent shot.[77]

In August 1859, Colonel Wilford told a class of Rifle Volunteers at Hythe that "in making a soldier a good shot you must raise his intelligence and in doing this you have a fair prospect of raising his character and making him a better man."[78] Lieutenant Walker agreed with Wilford, and had undoubtedly read Jacob as well. Instead of the "pipeclayed automaton of former days," Walker wrote of "the conversion of a mere human machine into a *taught* soldier [who are] not addressed as mere soldiers but as men endowed with an intelligent will, capable of development to a good purpose in the day of battle."[79] The result of such training is a soldier "who, in the field of active service, whether in the main body of an army, or separated from it as in skirmishing, would still, by virtue of his special training, elicit the full power and capabilities of his rifle."[80] An instructional companion to the new Enfield rifle-musket described it as "one of the most perfect weapons (if properly used) that science has yet placed in his [the soldier's] keeping."[81] To properly use the new weapon that represented an astonishing leap in military technology, the soldier had to think. Dougall boldly asserted "the Enfield rifle will prove the *intellectual man's weapon*" and clarifies

---

[77] Wilford, 13.
[78] King, V. A. *Three Lectures Delivered to the Second Company of Cheshire Rifle Volunteers* (Webb and Hunt, Liverpool:1859), 8
[79] Walker, Arthur, *The Rifle: Its Theory and Practice* (London: 1865), 11.
[80] Ibid., 10.
[81] Browne, S. Bertram, *A Companion to the New Rifle Musket* (London: 1859), 5.

himself, "that is, the soldiers must always be trained up to their weapon."[82] Another contributor to the *United Service Magazine* agreed with the advancement of the common soldier, citing the skill and training necessary to operate the new rifled weapons as "proofs that the soldier has risen, in the public and official esteem, to the lofty rank of a rational being."[83] In 1859, General Sir John Fox Burgoyne commented at the United Services Institution that it "is becoming more and more apparent, that the soldier's is becoming every day more a *skilled* profession, and the *well-practised* soldier of *increasing value* over the raw recruit."[84] Lieutenant Watson, in his 1862 lectures, rhetorically asked "*Can every man be taught to shoot? Can he be made a good shot?* Certainly he can."[85] The final word may belong to General Hay, the commander and chief defender of the School of Musketry, who put it plainly: "I am quite satisfied from what I have seen, that the private soldiers of the infantry are quite intelligent enough to acquire a sufficient knowledge, theoretically and practically, in the use of their rifle, so as to render them more than a match for any field artillery at present in existence."[86]

With the adoption of the rifle-musket, the value of the common infantryman was raised, and not just on the battlefield but in a broader context. The dividing line was the rifle. The common soldier rose from the pipeclayed unthinking automaton with a smoothbore to a valued professional with the rifle, a thinking and intelligent man, interested in his profession, a skilled artisan whose medium was gunpowder and lead. The editors of the *Saturday Review* noticed this phenomenon in 1860, in a review of Colonel Wilford's *Lectures Upon the Rifle*. The soldier "is not, as many eminent commanders have considered him, a mere

---

[82] Dougall, 31.
[83] *United Service Magazine and Naval and Military Journal*, Part 1, 1863, 434
[84] Wilford, 62. Italics in original.
[85] Watson, 8.
[86] Tyler, R. E., "The Rifle and the Spade, or the Future of Field Operations," *Journal of the Royal United Service Institution, Volume 3* (1860), 170

machine, but that he has a mind capable of understanding the theory upon which the routine of drill is founded." They go on to observe that if "the soldier's profession calls for the exercise of high intellectual qualities, a better class of men will be attracted into the ranks, and they must necessarily have officers of capacity not inferior to their own; and thus the whole army will gain in social consideration."[87] Not only was the value and social standing of the soldier increased, but also the effectiveness of the individual soldier who was now engaging his own target and operating on the battlefield with considerable independence. This seemed likely to supersede the tactical genius of academy-trained officers maneuvering bodies of crisply drilled formations. The old soldier General Burgoyne readily agreed: "It is evident," he said, "that the general introduction of the rifle into the service, and the greatly increased power that is rapidly being developed by means of rifled cannon, will require much alteration in the system of warfare."[88] In other words, the soldier had transitioned from the Napoleonic to the early modern. Although the rifle-musket was a muzzleloader with a slow rate of fire and parabolic trajectory, it was still a rifle.

Since musketry instruction was essential for the soldier to efficiently use the rifle-musket, institutions for providing this instruction were promptly opened. The first was the French artillery school at Vincennes, which began teaching the principles of rifle shooting as early as 1851; the tradition of artillery officers teaching rifle instruction began here. The Vincennes course was four months long. In 1853, as the Pattern 1851 rifle was being adopted by the British Army, Lord Hardinge established the School of Musketry at Hythe, on the Channel coast in Kent. By 1857, similar schools had appeared across Europe: in 1855, Spain opened a school at Madrid and Sweden at Stockholm, shortly after the Dutch established a school at Haye, and the Russian school was established at St. Petersburg in 1857. Conspicuously absent from the list of nations with established musketry schools

---

[87] *Saturday Review*, 55
[88] Wilford, 61

was the United States; the U.S. Army's School of Musketry at the Presidio of Monterey, in California, would not open *until 1906*, and then only to train soldiers on the new Maxim machine gun. Most of the continental schools (including Vincennes) were strictly for officers only, but at Hythe "both officers and privates are trained together, with obvious advantages."[89]

"Now musketry instruction is the means by which soldier is taught not only to shoot but in every way *make the most of his rifle*," Watson said in 1862.[90] In the British Army, a systematic method of training was established with the publication of the 1855 *Instruction of Musketry* regulation. Nearly all of *Instruction of Musketry* was written by Captain Lane-Fox, who was now Chief Instructor of the satellite musketry school at Malta and who had established his system of musketry training while at Malta and Varna in 1854, teaching the British troops bound for the Crimea. Lord Hardinge specifically charged the generals in command of the various military districts to use the *Instruction of Musketry* regulation and "make a point of ascertaining, at their periodical Inspections, that the Instructions [of Musketry], as therein laid down, are strictly observed by the troops under their orders."[91] Each battalion would send a few officers and NCOs to the School of Musketry at Hythe, who would return to their battalions as trained instructors. Colonel Wilford explained the process in 1859: "Detachments of different regiments are sent to that establishment [Hythe] to go through a course of training, which occupies about ten weeks, in order that each regiment may be supplied with a qualified officer and non-commissioned officer instructor in musketry."[92] A second, more comprehensive *Instruction of Musketry* was published in 1856. This, in turn, was superseded by the *Regulations for Conducting the Musketry Instruction*

---

[89] Walker, 22
[90] Watson, 8. Emphasis in the original.
[91] *Instruction of Musketry* (London: Military Library, Whitehall, 1855), circular memorandum frontispiece
[92] Wilford, 12.

of the Army, 1859. General Hay, the Commandant of the School of Musketry, appears to have had strong professional disagreements with Captain Lane-Fox over the methods of instruction and, probably, the scientific theories behind rifle shooting.[93] The 1859 regulation reflects General Hay's ideas, which differed slightly (but in essential areas) from Lane-Fox. Also in 1859, an additional chapter on "Musketry Instruction" was added to the *Field Exercises and Evolutions of Infantry,* the primary infantry manual for the British Army. Here was found the somewhat famous statement, often quoted with zeal by period authorities: "A soldier who cannot shoot is useless, and an encumbrance to the battalion."[94]

At the School of Musketry at Hythe, the officers and sergeants destined to become musketry instructors in their battalions received a number of lectures and practical exercises before firing a shot. Perhaps the best description of the School of Musketry, circa 1860, was written by Henry Edwards, a Rifle Volunteer who wrote a short narrative of his experiences at Hythe. "It is not a hap-hazard mode of teaching, to be applied in any fragmentary, fitful way, variable by individual tasks," Edwards wrote in 1860, "but a well devised, well tried, methodized system, to be taught systematically, as a complete whole."[95] The course at Hythe (and the British Army's musketry instruction in general) was designed to teach proper shooting fundamental skills *prior* to actually shooting live ammunition. Musketry instruction was separated into two parts, beginning with several weeks of Preliminary drills before the student proceeded to the Practice (or actual firing) final phase of the course. As Edwards explained in his narrative, "One of the great maxims of the school is this: We teach a man to shoot without

---

[93] Lane-Fox
[94] Field Exercise and Evolutions of Infantry (London: Horse Guards, 1859), 381
[95] Edwards, 11.

ball, and then we give him ball to prove what he has learned."[96] In his *Hand-Book for Hythe,* Hans Busk wrote that by strict adherence to the preliminary lectures and instruction, "far more than average proficiency in shooting is attainable without the expenditure of a single ball-cartridge."[97] Lieutenant Walker agreed: "provided a man thoroughly masters his preliminary drill, we undertake to make him a good shot before he fires a single bullet."[98] Colonel Wilford said "We use balls merely to ascertain if the men have profited by their teaching, or how they have been taught."[99]

The preliminary drill at Hythe consisted of a series of eight lectures and exercises, and assumed the students were utterly unfamiliar with firearms. A total of 52 distinct drills, each meticulously observed with any defects instantly corrected, were conducted over a period of several weeks before the student would fire the first live cartridge. The cadre at Hythe preferred students who had no firearms experience, and did not have any bad habits to painfully unlearn. According to Hans Busk, "The less practice he has previously had with the rifle, the better shot he is likely, in a limited period, to become."[100] A Rifle Volunteer recalled General Hay's initial lecture to them, at Hythe in 1859: "He told us that the less shooting we had previously had the better; for those of us who have never fired a gun were most probably the best, as he had merely to teach us to shoot, while those of us who had been in the habit of shooting were probably the worst, as he would have first to break us off bad habits."[101] In order, the eight subjects of the preliminary instruction were cleaning arms, theoretical principles, aiming drill, position drill, snapping caps, blank firing, judging distance drill, and the manufacture of cartridges. Of the eight, four were considered far

---

[96] Ibid., 18
[97] Busk, Hans, Hand-Book for Hythe (London: 1860), 57
[98] Walker, 13.
[99] Wilford, 29
[100] Busk, *Hand-Book*, 58
[101] King, 8

and away the most crucial: the theoretical principles, aiming drill, position drill, and judging distance.

In the theoretical principles lecture, soldiers were taught the basic elements of what is today called "ballistics." With the aid of chalkboards, diagrams, and models, the trajectory of the bullet at various distances was illustrated. The forces acting on the bullet were explained, and the soldier was introduced to technical terms such as the *line of sight, line of fire, trajectory,* etc. A secondary, but important, role of the theoretical principles lecture was to spark the interest of the soldier, and provide an explanation that would form an underpinning for the often-mundane tasks and drills encountered later in the course. "We teach first by theory, because a soldier is a being with a mind," Wilford said in an 1858 lecture, "and by theory we give him the *reasons* for everything he may afterwards be called upon to perform in practice." Essentially, Wilford concluded, "it is not enough that a man is able to hit his mark or miss it; he must know *why* he hits or misses."[102] The officer-instructor giving the theoretical principles lecture should, according to Walker, "convey to his hearers in a clear and concise manner a knowledge of the physical laws which regulate the flight of the bullet." Furthermore, the instructor must also "endeavor to interest the mind of the soldier in the subject of rifle-shooting generally, by supplying him with the reasons for all the details of his drill."[103]

Aiming drill was conducted by placing rifles atop a sandbag perched on a tripod of wooden poles, and having the soldiers carefully aim at a specified mark between 100 and 900 yards away. The musketry instructors would the check the aim, ensuring that the sights were level and the correctly set for the specified distances. Soldiers were familiarized with the operation of the ladder sight on the P1853 Enfield, a sight many critics claimed was too "delicate" for common soldiers. Generally speaking, however, the sights on the P1853 rifle proved robust; out of thousands of soldiers that had passed through the School of

---

[102] Ibid., 31
[103] Walker, 15

Musketry at Hythe, Colonel Wilford recalled that there had only been one soldier who managed to break or damage the back sight of his rifle.

This was followed by the position drill; depending on which 19th century authority you asked, it was either the position drill or judging distance drill that was considered the most important. Today, most of the position drill would be considered the basic good shooting fundamentals, including breath control, sight picture, and trigger squeeze. The key to the position drill was consistency, doing the same thing every time. In this drill, soldiers were taught how to hold their rifles, where to grip the stock, and the crucial concept of a smooth trigger squeeze. The British Army issued a device called a *snap cap* to soldiers, consisting of an iron bushing that fit over the rifle's nipple, topped by a leather pad. This allowed the soldier to "dry fire" his rifle without injuring the nipple, and in this way, soldiers could practice their trigger squeeze. Spots were painted on the walls of barracks, so that at various times throughout the day, soldiers could take up their rifles and practice the position drill, aiming at the spots. Soldiers aspired to be so steady in their aim and trigger pull that, according to Hans Busk, a "penny piece" could be balanced on the musket barrel for the entire position drill, without falling.[104] The 1859 *Field Exercise and Evolutions of Infantry* called it an "important drill, of which there cannot be too much, if well executed."

Another essential component of the position drill was to develop strength in the upper body muscles, so that the exercise of loading and bringing up to the shoulder the heavy rifle-musket (which weighed 9 pounds 12 ounces with bayonet fixed) would not cause fatigue. These many repetitious exercises made position drill by far the longest of all the drills, with four consecutive days of the position drill practiced morning and afternoon. This quickly made the position drill the soldiers' least favorite (but the longest by far) of all the musketry drills. The degree of importance placed on the position drill is summarized in the

---

[104] Busk, *Hand-Book*, 63.

circular "General Order" at the beginning of the 1859 *Regulations for Conducting the Musketry Instruction of the Army*:

> With a view to compensate for the short time devoted to the execution of the preliminary drill *as part of the yearly course,* the General Commanding-in-chief desires that the position drill, particularly the first and third practices, may be frequently performed under close supervision, in order, by strengthening the left arm, to give the soldier a perfect command of the rifle with his left hand, and to establish that union between hand and eye, which is indispensable for good rifle shooting, but which can only be attained by constant practice. By command of His Royal Highness, the General Commanding-in-chief.[105]

The final drill of the preliminary instruction was judging distance. This was a crucial skill. As the range to the target increased, the *dangerous space* correspondingly decreased. This is the great weakness of the rifle-musket; to hit at ranges beyond about 400 yards, the great arching trajectory of the bullet requires the soldier to determine the distance to his target with some degree of precision. It has already been demonstrated that for shooting at 600 yards, a distance estimation only has to be 10 or 20 yards in error to cause the bullet to miss a man-sized target entirely. At 900 yards, the dangerous space is narrowed to a mere ten yards and an error in distance estimation of scarcely 5 yards (or 0.5%) would result in a miss. At these ranges the bullets were almost literally falling from the sky, dropping 6 feet in only 10 yards; as Wilford described, it has "to be dropped, as it were, almost *on* the target, and which of course is very difficult."[106] It is primarily on this point that modern historians (usually of the American Civil War) dismiss the rifle-musket as a revolutionary arm, and certainly reject it as the first modern infantry weapon.

---

[105] *Regulations for Conducting the Musketry Instruction of the Army*, 4.
[106] Wilford, 50

To paraphrase the general consensus of contemporary historians writing about the rifle-musket (like Dr. Guelzo, Dr. Hess, et al.), the trajectory made the weapon too difficult to use at anything but close range, and relegated its tactical value to little better than (or even inferior to) the smoothbore.

The British military authorities in the mid-19th century fully understood the difficulty of accurate long-range fire, but they were determined to exploit the greatest possible capability out of the new rifle. As Hans Busk explained, "It is not sufficient to be able to strike, even with tolerable certainty, a target stationed at a known distance." The soldier had to judge the distance rapidly, and with reasonable accuracy, even in combat. "An error of a few yards in this respect, will render the best rifle valueless in the hour of need," Busk admitted without hesitation. "If the rifleman do not therefore possess in an eminent degree, the knack of correctly judging distances, he is of comparatively no use in the field. That this power is attainable by practice, every day's experience, more especially at Hythe, conclusively proves."[107] Even Colonel Wilford freely acknowledged the rifle-muskets rainbow trajectory that made shooting at longer distances notoriously difficult. "If, with the Enfield rifle," Wilford wrote, "if I can hit a target half a dozen times running at 600 yards, and strike it but once out of six at 900 yards, I am equally satisfied." Colonel Wilford understood the problem, and hoped someday the British Army would be equipped with a futuristic rifle, as then yet unknown, with a flatter trajectory. "Give me a rifle much lower in its trajectory," Wilford said longingly, "when it [the target] could be struck as easily at 900 yards as it is at 600 yards."[108] In his treatise, Walker acknowledged "most of our difficulty in shooting is owing to the bullet moving in a curved line."[109] He observed that, when fired at a 900 yard target, "the Enfield bullet actually acquires a height of 50 feet when culminating," so that on it's way down,

---

[107] Busk, *Hand-Book*, 95
[108] Wilford, 50
[109] Walker, 142

> It may be said to approach the vertical; and herein lies the difficulty of obtaining accurate shooting with the Enfield at this and other long ranges; we have, as it were, not only to fire *at*, but *into* an object; the bullet has to be dropped on a man's shako in order to hit him; so that the slightest possible error in the matter of elevation or taking aim is at once fatal: a finer sight than usual will cause the bullet to fall short of the mark, while a fuller sight will cause it to fall over and beyond.[110]

The difficulty was understood; the answer, provided at the School of Musketry and in the musketry instruction given to the British Army, was extensive practice at judging distance and modest expectations of actual accuracy at long distances. "Before an enemy the distance is unknown," the *Instruction of Musketry* regulation of 1855 (quoted verbatim by the 1859 *Regulations*) said, "it is therefore necessary, in order to apply the rules laid down for shooting, that he [the soldier] should know how to judge quickly, and with tolerable accuracy, the distance which separates him from the object he is firing at, so as to regulate the elevation of the rifle." In the course of the judging distance drill, soldiers were posted at various distances from the students, starting at 50 yards and thence every 50 yards thereafter, e.g. 100 yards, 150 yards, 200 yards, and so on, out to 600 yards. The student was trained to carefully observe a man at each of the set distances, "and register in his mind all the particulars concerning his appearance."[111] At 50 yards, for instance, a soldier with average eyesight could name any man in his regiment, and details of uniform and complexion are visible. By 200 yards, however, the buttons of the coat are invisible and the face is a blur, while larger badges on caps and cartridge boxes can be discerned. Out to 600 yards, infantry appear to have no eyes or necks, the head is a ball

---

[110] Walker, 140
[111] Wilford, 40

upon the shoulders, and flesh is not visible. At 800 yards, the movement the arms and legs of marching soldiers can just be seen. Soldiers were expected to memorize how men appeared to them at these various distances, so that a thoroughly trained soldier could instantly and automatically estimate the distance to enemy soldiers on the battlefield. Colonel Wilford likened it to the effortless skill at range estimation that English longbowmen developed over a lifetime of training.

It was important that every soldier know how to judge distances. The 1859 regulations conservatively instructed soldiers to set their sights to distances provided by their officers, but if a range was not given, the soldier was expected to estimate his own range and set his sights accordingly. The emphasis at the School of Musketry was for each man to estimate distance himself. This meant that in a battalion of a thousand riflemen, there would be a thousand estimations of the range. Some (perhaps even a great many) estimations would be incorrect, and increasingly so out to longer distances. Some, however, would be correct. Ultimately, some rounds would be on target. The longer the distances, the lower the percentage of hits, but this was unflinchingly accepted by the Hythe authorities. They were not frustrated by the fact that, at 900 yards, the overwhelming majority of the bullets fired would miss. Instead, they appreciated the astonishing fact that a small percentage of bullets would likely *hit*. Russian columns and batteries learned this lesson the hard way in the Crimea, and after Inkerman, they never again dared to venture beyond their defensive works and engage the allied armies in the field. None of this would be possible with smoothbore muskets, or with untrained soldiers. A large target like an enemy column, 900 yards away, could be effectively engaged by soldiers with a rifle like the Pattern 1853 Enfield, even if it would be virtually impossible to hit an individual enemy soldier at that range. An article in *The Saturday Review* in 1860 clearly explained this tactical concept, no doubt with the assault columns of the French in mind as the "hostile battalion" advancing upon British troops:

To drop a ball from an Enfield rifle on to a reconnoitering staff officer at 900 yards could be done probably by only a few men out of a battalion. But if a hostile battalion should endeavor to advance from 900 yards distance to 300, every rifleman opposed to them ought to do a considerable part towards their destruction. There would be no need of nicely timing the discharge, or of minute accuracy of elevation. A ball which will fly for 100 or 200 yards within five feet of the ground must necessarily do some harm if it commences that part of its course somewhere near the front of a body of advancing troops.[112]

Both at Hythe, and in the annual course of musketry instruction prescribed for every soldier in the British Army, actual shooting of live cartridge did not start until the preliminary drills had been completed. The 1859 *Regulation* made it very clear that soldiers "are on no account to be permitted to fire ball until they have been exercised in all the subjects" of the preliminary drill.[113] When the last of the preliminary drills were complete (and concluding with judging distance practice), the soldier began firing his allowance of ammunition in an annual target practice. Although it was called "target practice" in the regulation, it was, in fact, an annual rifle qualification. Regiments and battalions competed to achieve the highest overall "figure of merit," a Victorian measure of shooting accuracy. For individual soldiers, they were scored and graded on rifle shooting as well as judging distance; it took good shooting, and above average skill at judging distance, to enter the first-class. Every year, the best shot in the battalion was awarded a golden badge and an extra two pence in pay per day. Up to 100 soldiers in each battalion, who had scored in the first-class in judging distance and shooting, received a smaller badge and a penny per day.[114] For privates being paid

---

[112] "Lectures on the Rifle." *The Saturday Review,* January 14, 1860, 55.
[113] *Regulations*, 59
[114] Ibid., 72

nominally one shilling (12 pence) a day before stoppages (of up to sixpence per day) were deducted, the shooting prizes represented a substantial increase of their overall income.

Experienced soldiers would fire 90 rounds, at targets placed 150 yards to 900 yards away from the firing line. New recruits and officer cadets fired 20 rounds in "preliminary ball practice" before firing 90 more rounds at targets from 150 to 600 yards. The targets, by modern standards, were generous; the expectations of accuracy achieved by line infantry soldiers were rather modest even for their day. From 150 to 300 yards, the soldiers shot at a target four feet wide and six feet tall, or about the size of two men standing side by side. This was increased to a target eight feet wide and six feet tall from 400 to 600 yards, and the 650 to 900 yard target was 12 feet long but still only six feet tall. These targets represented lines of enemy soldiers, not individuals. A hit *anywhere* on these rather enormous targets counted for a point, with additional points scored for hits in much smaller bulls-eyes. The size of the targets, being much larger than a man-sized individual, reflect a very practical approach to military musketry: precise bulls-eye shooting was excellent and rewarded, but putting rounds in the general proximity of the enemy at ranges over 300 yards was sufficient for battlefield purposes. In an era when infantry moved in large formations, the formations themselves became the target. It didn't matter if a soldier couldn't hit an individual enemy at 600 yards; what mattered was if he send a bullet near or into the enemy formation, or enemy battery, at 600 yards. Colonel Wilford recalled that 30 officers "who had never had any instruction in rifle shooting" arrived at Hythe on 15 August of 1859; they were each given three rounds and directed to fire at a 600 yard target. Only 18 rounds out of 90 hit the target. The officers were quickly given a crash course in musketry instruction, and they tried again. This time 30 out of 90 rounds hit the target, "thus nearly doubling their efficiency, as the consequence of being taught to shoot."[115] The fact that two-thirds of the rounds

---

[115] Wilford, 65

missed entirely didn't bother Colonel Wilford or the cadre at the School of Musketry, who appreciated the fact that one-third of the rounds *hit*.

The expenditure of 90 (or 110) rounds per year, per soldier, in rifle qualification is also remarkable for an army in the 1850s. It amounted to the expense of millions of cartridges annually, at the cost of tens of thousands of pounds. Tens of thousands of regular army soldiers and thousands more in the militia and volunteers shot the annual "target practice," and the *majority* entered the first-class. From 1856 to 1858, over 400 officers and 2000 sergeants and privates were trained at the School of Musketry at Hythe; 84% of the officers, and 79% of the enlisted soldiers, qualified in the first class.[116] These trained officers and men returned to their regiments as musketry instructors. Their knowledge percolated through the British Army and, year by year, the number of Hythe-trained instructors continued to increase. In 1861, a second school of musketry opened at Fleetwood to further increase the density of trained instructors in the ranks. The British Army was unique as a professional army, with most soldiers enlisting for long periods of service. They were soldiers by profession, the army was their sole career. Year after year, each soldier received the mandated musketry instruction, and was then tested in the annual qualification on his ability to judge distances and to accurately fire his rifle at ranges from 150 to 900 yards. There is sufficient evidence to say this training program was in fact *modern marksmanship instruction*, given to soldiers with the first modern infantry weapon. There was nothing like it previously, and it remains fundamentally the same sort of training given to soldiers today, who still use a rifle as the basic infantry weapon.

---

[116] Ibid., 45. The musketry instruction regulation changed in 1859, condensing the course at Hythe; the percentage of soldiers graduating in the first-class dropped notably, to about 67%. In addition to the course changes, instruments were used to determine the distances in the judging distance drill, instead of the old Gunther's chain. Apparently, according to Hans Busk in the *Hand-Book*, page 164, students had found ways to cheat because of the chain's "assistance."

The instruction given to British Army recruits carrying P1853 Enfield rifles in 1859 is strikingly similar to the training given to United States Army recruits carrying modern rifles today.

Not many years ago, I was brand-new recruit with a freshly buzzed head in Basic Combat Training on Sand Hill, at beautiful Fort Benning, Georgia. About two weeks into "boot camp" my company started a period of classroom-based training on "Ballistics and Zeroing," which is the first block of instruction for Basic Rifle Marksmanship (BRM). Like nearly all classes in the early 21st century U.S. Army, this one was on PowerPoint. With failing patience betrayed by an unending stream of expletives, our drill sergeant did his best to teach a platoon of hungry, sleepy teenagers about such concepts as line of sight, line of bore (known to British recruits circa 1859 as *line of fire*), and how gravity and air resistance influences the trajectory of our M16A4 rifles. For weeks, we practiced our marksmanship at every spare moment with unloaded weapons, including a "Dime/Washer drill" for developing a smooth trigger pull; Hans Busk would have recognized the Dime/Washer Drill as the penny piece exercise he wrote about in 1860. I don't think Colonel Wilford would have fully approved the hasty marksmanship instruction we received in basic training, but he would have been absolutely delighted by the worn-out M16A4 rifles we carried. Built with a "battle sight zero," these rifles can hit a target "center mass" from 25 out to 300 meters, without any adjustment of the sights for range. It is simply point and shoot; compared to the bullet from a rifle-musket, the M16 rifle is practically a laser pointer. Over 300 yards distance, the 5.56mm bullet from an M16 rifle drops only a few inches, compared to eight or nine *feet* of bullet drop with a rifle-musket. This was the rifle Colonel Wilford dearly wanted; he once wrote "let the manufacturers of fire-arms devote all their energies to improve the soldier's rifle by lowering the trajectory, and thus make him a more deadly adversary to the Queen's enemies."[117] He died in 1880 at the age of 81, just a few years before France adopted the

---

[117] Wilford, 43.

flat-trajectory Lebel rifle with smokeless powder cartridge in 1886.

The modern military rifle training is, however, essentially the *same instruction* given to British Army recruits in the 1850s, and surprisingly little changed across the Atlantic and across more than 15 decades. Some terms had been changed, and some of the drills has been amended or done away with completely thanks to a much flatter trajectory, but overall it was the same. Technology had improved the rifle, but it was still a rifle: both the M16A4 that I carried in basic training, and the P1853 rifle that a British Army recruit carried during his depot training, could reliably hit targets at 300 yards after some brief instruction. No smoothbore could do that, no matter how precisely the gun was built or how thoroughly the soldier had been trained in the use of his musket. The inverse was also true: without training, a brand new inexperienced recruit with either an M16 or P1853 Enfield rifle would be unlikely to hit a target at 300 yards.

7
SUPPORT BY FIRE

While much attention was being given to training up the British soldier in the use of the P1853 rifle in the last years of the 1850s, specific refinements were being made to the ammunition it used to facilitate superior battlefield performance. Originally the Enfield cartridge utilized a .568-caliber bullet designed by the London gunmaker Robert T. Pritchett. The bullet was wrapped the cartridge paper that had been dipped in a lubricating composition (of tallow and beeswax), and the bullet was rammed down the barrel while still enveloped in the cartridge paper. The cartridge performed well in testing and experiments at Hythe, but service in the Crimean War revealed that the Pritchett bullet expanded slowly and caused rapid fouling. In 1856, a wooden plug was added to the bullet to ensure rapid expansion; the rest of the cartridge remained unchanged. During the Indian Rebellion of 1857, this bullet also became difficult to load after the rifle had been fired several times in harsh battlefield conditions.

The decisive improvement came in 1859, when Captain Boxer of the Royal Laboratory at Woolwich reduced the diameter of the bullet from .568-caliber to .550-caliber.[118] It retained the wooden plug, which remarkably forced a .550-caliber bullet to expand into the rifling of a .577-caliber barrel. This new cartridge, with the .550-caliber bullet, was just as accurate as the .568-caliber bullet and had a slightly flatter trajectory. More importantly, it could be loaded with ease and fired dozens and even hundreds of times without difficulty. With this perfected ammunition, the rifle-musket reached the peak of efficiency. Contemporary cartridges,

---

[118] Captain Edward M. Boxer is best known today as the inventor of the Boxer primer system for centerfire rifle cartridges, still in widespread use as the most common primer system for small arms cartridges.

such as the type used in the US Army's M1855 and M1861 Springfield rifles, were capable of accurate shooting but lacked the lubricant-coated paper patch the Enfield cartridge employed. This resulted in rifles rapidly becoming fouled and difficult to load, often after firing a relatively few number of rounds. With Enfield cartridges, on the other hand, British soldiers could keep up an accurate and sustained fire, as long as the ammunition supply held out. To prove a point, an Enfield rifle at the School of Musketry was fired once every day, over a period of several years, without cleaning out the barrel between shots.

The adoption of the rifle by trained soldiers also led to a subtle yet significant change in the British Army's infantry drill manual. In the old smoothbore army, the firing commands for volley fire were given by officers. The firing procedure began simply with "Make Ready" at which time the soldier would bring the musket hammer to full-cock, followed by the command "Present." At the command to *present*, the soldier raised the musket to his shoulder, awaiting the command of "Fire" at which instant he sharply pulled the trigger. While the older drill manuals during the Brown Bess era urged soldiers to aim down the top of the barrel of their muskets, it was (as Colonel Wilford and others have lamented) rarely, if ever, actually done in practice. In the 1850s, the *Field Exercise and Evolutions of Infantry* was updated; gone was the command "Fire" that ordered the soldier to pull the trigger immediately. Instead, the new sequence of firing commands began with the officer in charge providing the range that he had estimated for the distance to the target. For instance, the officer would give an order such as "At 400 yards, Ready!" The soldiers would set their sights to 400 yards, bring the rifle hammer to full-cock, and wait for the command of "Present." Upon this order, the soldier brought the rifle up to the shoulder and was allowed a period of several seconds to aim; when he had a good sight picture and was ready to shoot, the soldier carefully squeezed the trigger. Such a "volley" would sound somewhat ragged, as soldiers fired their rifles on their own time, over a period of a few seconds. But the destructive power of such a volley, with *aimed* fire, was considerably greater than the crisp volleys delivered under the old regulation, which demanded every soldier pull their triggers at the instant of hearing the command of "Fire" without any particular concern for how well the soldier was aiming.

Combining a rifle capable of accurate fire with ammunition that allowed accurate and sustained fire opened up new tactical possibilities. Today in the US Army, we would call the P1853 Enfield rifle with its painstakingly perfected ammunition a "combat multiplier," i.e. it gave the average trained British soldier an advantage over the average soldier of any other army equipped with an ordinary rifle-musket. While many British officers were true to the stereotypes and dug in their heels at any thought of change, a growing number of British military authorities were aggressively seeking to identify new force multipliers for the British Army, particularly in the areas of training and weapons technology. It is probably not a coincidence that the search for, and implementation of, various force multipliers for the British Army was happening amid the background context of a major war scare with France, at the end of the 1850s.

Great Britain and France were unlikely allies in the Crimean War. Among the many lessons the British Army painfully learned from the Crimean experience was that France could do war much better than Britain. Amid the finger-pointing and blame-shifting of the disastrous winter of 1854-55 and the deteriorating performance of the replacement-filled units at the assaults at the end of the Sevastopol siege, the British realized with a sinking feeling that their army had atrophied terribly from the once-powerful victors of Waterloo. The French deployed a much larger number of troops to the Crimea than the British, and managed to adequately equip and feed them while the much smaller British contingent froze, starved, and died of preventable diseases. In nearly all aspects of logistics and medical support, the French demonstrated far superior capabilities. It was immediately and soberly realized that if France could project a large army hundreds of miles to the Crimea with such effect, then the French could easily project an even larger army across thirty miles of English Channel.

Fortunately, Britannia still ruled the waves, and the French would have to deal with the Royal Navy before they could seriously contemplate an invasion. With hearts of oak, the men and ships of the Royal Navy would smash any French invasion fleet to splinters before it could reach England's green and pleasant land. But another paradigm shifted in 1858, when France laid down *La Gloire,* the world's first oceangoing ironclad warship. With armor plate nearly five inches thick, *La Gloire* was impervious to the guns of the Royal Navy. Protected by this unstoppable ironclad monster, a French invasion fleet could cross the Channel. Two more ironclads were quickly laid down in succession at French dockyards, and for a time, a successful French invasion in the event of war seemed a very real possibility. The British press sounded the alarm with all the verbose hysteria of mid-19th century Victorian journalism.

Relations between the powers plunged in 1859, as *La Gloire* and her sister ships were launched. At the same time, France had intervened in the Italian wars of independence, aligning with Sardinia against Austria and rocking the precarious diplomatic boat. In Ian Beckett's book *Riflemen Form,* his survey of just the *Saturday Review* found no less than 15 "war-scare" articles in 1858, 47 in 1859, 44 in 1860, 33 in 1861, and even 25 in 1862.[119] Tennyson's alarmist poem, *The War,* was published in *The Times* in May, 1859, urging Rifle Volunteers to form "in freedom's name" for the defense of the Queen; Napoleon III was the obvious antagonist, described as the "despot." Invasion was seen as possible, perhaps even imminent. Books about rifles and rifle shooting became instantly popular, and one of the greatest challenges of writing this book was simply deciding, out of the glut of British works on rifles in the 1850s, which ones to cite as the best examples of the genre. The *rifle* of the British soldier and

---

[119] Beckett, Ian F. W., *Riflemen Form: A Study of the Rifle Volunteer Movement 1859-1908*

Volunteer was to be England's defense.[120] This was the background day to day context of the British Army in the late 1850s, as the School of Musketry (which soon opened the satellite campus in Fleetwood, in 1861) conducted its essential training with a real sense of urgency. The instructors, and the students, truly believed their training would be put to the test in actual combat with a numerically superior professional enemy, at any moment. The rifle in the hands of a relatively small but well-trained body of fighting men was to be England's force multiplier against superior numbers. Immense faith was placed in the rifle, but it was faith primarily based on practical experience and not a mere zealous hope. Hard fighting in the Crimea and India had demonstrated that the rifle, in trained hands, could certainly be a combat multiplier; in these cases, however, the enemies were usually armed with smoothbore muskets. Just what such an Anglo-French War of 1859 would have looked like is, and how effective the Rifle Volunteers would have been in support of the British Army in the event of French invasion, can, of course, only be left to speculation.

By about 1858 the British Army had completely adopted the Enfield rifle as the general arm for the entire service, for both line and light infantry alike. The rifle itself was inherently accurate enough to hit individual man-sized targets out to perhaps 600 yards, and "area" sized targets to 900 yards and beyond. Finally, the ammunition for the Enfield rifle had been incrementally improved upon, year by year; the final perfected cartridge gave the British soldier the ability to maintain a sustained, indefinite fire as long as the soldier had a supply of ammunition. With the French "war scare" providing the immediate and urgent motivation to seek out any and every possible battlefield advantage and combat multiplier, British strategists and

---

[120] While the Rifle Volunteers found a number of enthusiastic supporters among the regular officers of the British Army, the movement as a whole was seen, by and large, by the regular army as a good patriotic expression, not to be discouraged, but which would be of uncertain value, at best, in the actual event of war.

proponents of the rifle-musket began the first incipient foray into what today we call modern infantry small unit tactics. In this climate, the instructors at the School of Musketry realized the potential of what they had to work with, and began considering the broader capabilities of the rifle-musket with the perfected Enfield cartridge. As the "storm of battle and thunder of war" seemed to roll England's way, as Tennyson put it in *The War*, they asked the crucial questions: how can this weapon system be used in broader contexts, and what new or modified tactical options does this weapon system provide on the battlefield?

In a lecture given to the United Service Institution in May, 1858, while describing the "grand requirements" for the soldier's weapon, Colonel Wilford placed "celerity of loading" and "the power of firing a great number of rounds in a given time, combined with accuracy at long range" above all others.[121] Already, the School of Musketry had adopted comparatively enormous targets for the annual musketry qualification, representing "area" targets. Various experiments were regularly undertaken, with one such experiment consisting of an infantry element firing upon mock artillery batteries complete with life-sized stuffed targets representing men, horses, guns, and carriages at 600 and 815 yards (the infantry were not told the range and had to estimate the distance themselves). The period sources (particularly Wilford and Busk) are quite pleased with the results, even though only a small percentage of bullets fired actually hit a mock target. All the key pieces were in place, and it took surprisingly little time before British officers were discussing an entirely new concept in modern warfare: infantry providing fire support for infantry.

The experience of the Crimean War and the Indian Rebellion had taught the British officers at the School of Musketry that infantry alone, properly trained and equipped, and in the right circumstances, were able to silence artillery and defeat cavalry. It was a small but extraordinary leap to begin theorizing that infantry can provide its own fire support, without relying on

---

[121] Wilford, 49

support from artillery (or cavalry), in the attack. Previously, artillery was the sole source of supporting fire for attacking infantry. The storming of San Sebastian in 1813 saw a crucial innovation emerge, as the British artillery fired over the heads of the assaulting infantry, giving the infantry fire support *during* their attack. After an initial unsupported infantry attack failed, the British artillery 700 to 800 yards away opened a heavy fire upon the strong points of the French defenses for twenty minutes. The artillery had weakened the French positions, and the infantry resumed the attack and carried the city without the loss of a single British soldier to the overhead artillery fire. Forty years later, the common British infantryman was armed with a rifle that had the same range as the British artillery at San Sebastian, and was capable of sustained accurate fire. Thinking through these new possibilities in the late 1850s, several British officers theorized that in the absence of artillery, infantry could provide fire support to another attacking group of infantry.

The earliest mention I have yet found of the concept of infantry providing fire support for attacking infantry was by Colonel Wilford, in a lecture at the United Service Institution in London, on July 10, 1857. There, he described, in embryonic form, the "fire and maneuver" infantry tactics that would ultimately be developed only in the 20th century with the advent of automatic weapons. The topic of his lecture was the necessity of adopting the rifle as the general weapon for the British service, which was still a subject of some controversy. Eventually, making several points in succession, he arrived at the impact of the new rifle-musket on warfare. Wilford acknowledged that only "actual and varied warfare can alone demonstrate" the actual effectiveness and result of infantry armed with the rifle, but he emphasized nonetheless that "the whole system of tactics and fortifications must undergo important modifications." He listed several likely scenarios and situations, such as the bivouacking of armies at a more "respectful distance," and then casually mentioning mid-sentence, "infantry [will be] rendered less dependent for support from artillery." Almost as an aside,

Wilford elaborates: "A taught regiment of 800 men could throw 16,000 bullets in ten minutes into a fort of an area of 50 square yards, at a distance of 900 yards; and this could be done over the heads of a column advancing to storm."[122]

What Colonel Wilford described, in 1857, are the essential elements of the most basic infantry attack drill, developed to maturity in the First World War and still very much in use today. In what the US Army calls "Battle Drill 1" the unit is split into two elements: the assaulting element and the support by fire (SBF) element. In the most basic execution of the drill, the SBF element pours fire into the enemy position, fixing the enemy in place and forcing them to take cover; this enables the assaulting element to move forward in relative safety and attack decisively into the enemy's position. The success of the attack depends on the effectiveness of the *suppression* provided by the support by fire element. In his aptly-named and succinct article *Suppression is the Critical Infantry Task*, Major Brendan McBreen, USMC, explains that the assaulting element cannot drive home its attack until the enemy has been "*effectively suppressed*. This is the critical task. Effective suppression is a prerequisite for the assault and, in turn, the entire attack."[123] It requires the expenditure of a considerable quantity of ammunition, fired at the area where the enemy is *at* and not necessarily at a visible form of enemy soldiers themselves. Since Erwin Rommel employed mature fire and maneuver tactics on the Italian and Romanian fronts of the First World War, the SBF element has utilized machine guns to lay down effective suppressing fire upon the enemy position. Ideally, several machine guns are used, each firing in bursts to avoid burning out barrels, while ensuring the enemy position remains under a continuous, unrelenting fire. *Fire superiority* is achieved when the enemy is unable to emerge from cover to return effective fire. Done correctly, the SBF element allows the assaulting element to move safely into close proximity of the

---

[122] Wilford, 15
[123] McBreen, Brendan, "Suppression is the Critical Infantry Task." *Marine Corps Gazette* 10 (October 2001): 40. Emphasis in original.

enemy, so that when the SBF element lifts its fire, the hapless enemy raises their heads only to find an attacking force with all the initiative, bearing down directly upon them.

It is possible, perhaps, for the historian-exegete to tease out more of Colonel Wilford's brief comments. He was already trying to pack a great deal into the period allotted for his lecture, and the entire address feels somewhat rushed. While Colonel Wilford had several critics who disagreed with the methods he used at the School of Musketry and his ideas on how the rifle-musket was transforming warfare, even his critics readily agreed that Colonel Wilford approached musketry with the fervent passion of a true believer. Personally convinced that common soldiers could be trained to use a rifle effectively at long ranges, his lectures have the energy and conviction of the prophet standing on the mountain-top, proclaiming the truth to any who would listen. Colonel Wilford's brief, casual mention of fire and maneuver was given in the middle of a list he was rattling off, very quickly and with little further detail, of other tactical changes brought by the rifle-musket. If he was introducing the new and relatively complex concept of fire and maneuver for the first time, it may be reasonable to assume he would not have simply mentioned it in passing, in the middle of a list of other thoughts. This suggests that the concept of an infantry base of fire element supporting an attack element predates even the 1857 lecture.

Regardless of when exactly the concept was first proposed, it almost certainly was imagined as a hybrid combination of the comparable range of the rifle-musket with field artillery pieces, and linked the historic example of the British artillery at the siege of San Sebastián, in 1813. The remarkable and innovative use of artillery at San Sebastián would have been well known among the officers of the British Army some 40 years later, in the mid-1850s, especially among officers of the Royal Artillery. At this siege, the French had made a determined resistance. A practicable breach had been made by the constant battering of the British artillery and, after several disastrous earlier attempts, the British made a new assault with infantry on August 31, 1813. From the parapet and the top of the breach, the French poured a highly effective fire that cut down the British in swaths; unless something was done to check this fire, the assault was sure to fail. "It was at this time," the historian of the Royal Artillery Regiment wrote in 1879, "that Sir Thomas Graham ordered the artillery to commence a fire, which has received the greatest praise at the hands of historians." From a range of 600 to 700 yards, the British artillery opened fire at the French defenders, "over the heads of our own men (*only a few feet perpendicular lower down*) with a vigour and accuracy probably unprecedented in the annals of artillery." The new tactic worked, and for twenty minutes the British guns directed a careful and precise fire upon the French defenders, who continued "firing over the parapet as best they could, notwithstanding numbers had their heads taken off by our round shot."[124] The British soon achieved fire superiority, and the French fire slackened off amid explosions of French ordnance on the parapet. Then the British artillery lifted their fire, and the British infantry rose up and successfully stormed the breach, and took the city.

The successful storming of San Sebastián was well remembered. The commander of the British artillery at San Sebastián was promoted to command Wellington's artillery at Waterloo, and went on to retire a major-general. For a British officer like Colonel Wilford, a firm believer in the capabilities of the rifle-musket, simply replacing artillery with infantry providing covering fire for an attacking force is, perhaps, not too big a theoretical leap. Using rifle-armed infantry as a battery of sorts was already in the literature; in the 1856 *Pictorial History of the Russian War,* already quoted, Dodd wrote, "It was found that a rifle-pit was in effect a battery which, at 600 or 700 yards distance from a battery of large ordnance, had the power to drive away the artillerymen from the embrasures at which the guns appeared."[125] Instead of simply picking off gunners at the embrasures, Wilford

---

[124] Duncan, Francis, *History of the Royal Regiment of Artillery* (John Murray, London: 1879), 369
[125] Dodd, 404

expanded the target to a larger enemy-occupied area and added an assaulting element to move forward beneath the bullets of the base of fire element.

This concept of infantry self-supporting infantry was added to Colonel Wilford's lectures at the School of Musketry (and dutifully adopted by the lectures of subordinate instructors). In Harry Edward's Rifle Volunteer *Narrative* of 1860, he recounts one of Wilford's lectures about the Battle of Alma. "At the battle of the Alma the French essayed to turn the enemy's flank," Wilford explained, while:

> The English went right ahead at the batteries and were exposed to a murderous fire. They should have had riflemen behind them to quiet those batteries. At the storming of St Sebastian, the infantry advanced, the artillery firing overhead; and that must be our plan; put your riflemen behind, -- their ball rises 45 feet high. If you had only 800 men, they could fire three times in a minute; so in ten minutes they could pour in 24,000 balls -- such a shower of deadly hail as no living thing could stand; and such is the precision attainable, that if at 900 yards you laid down a table-cloth 50 yards square, I would engage with a company of taught soldiers to fill it with balls.[126]

From 1859 to 1860, Wilford's scenario had increased its rate of fire from two rounds to three per minute, perhaps attributable to the adoption in 1859 of Captain Boxer's .550-caliber bullet that could be loaded with ease, and fired almost indefinitely. Thirty rounds fired in succession would pose no difficulty whatsoever to the British soldier with a P1853 Enfield and the new .550-caliber ammunition. In this scenario, an average of 40 bullets *per second* would be falling into the target area, a truly astonishing volume of fire that is more than sufficient, even by modern standards, to suppress an enemy. Assuming a rapid sustained rate

---

[126] Edwards, 17

of fire of 200 rounds per minute (with frequent barrel changes) for a modern machine gun like the M240B or GPMG, twelve machine guns would be required to match Wilford's time and volume of projectiles on target.

The fullest expression of fire and movement with the P1853 rifle (complete with helpful illustrations) is found in Lieutenant Arthur Walker's *The Rifle: Its Theory and Practice,* published in 1864. At this time, Walker was an officer instructor at the recently opened satellite musketry school at Fleetwood, but he had spent some enough time at Hythe to become a devoted disciple of Colonel Wilford, whom Walker quotes liberally in *The Rifle.* Like Wilford, Lieutenant Walker recognized the rainbow-like trajectory of the Enfield rifle was a decided weakness, but believed this made it ideal for providing overhead fire. "Although the highly curved trajectory of the Enfield rifle at long ranges must obviously be regarded as a decided fault," Walker wrote, "yet certain exceptional circumstances may arise, under which this very fault may be turned to positive advantage."[127]

Lieutenant Walker then proceeds to describe a scenario wherein three 600-man regiments of infantry are required to attack an enemy fortified position on the far side of a stream. The hypothetical enemy very astutely had built their fortified work to cover the bridge over the stream; any attacking force, canalized onto the bridge, would be subjected to the direct fire from the fortified position. In ordinary circumstances, Walker noted, artillery would provide the covering fire to weaken the enemy position while the infantry advanced. "This was done, for example, with admiral result, at the siege of San Sebastian in Spain," Walker observed, clearly recalling Colonel Wilford's lectures. Yet for this scenario, the artillery was unable to come to the front, due to any number of possible reasons. Walker was not deterred: "Let us see whether, under such circumstances, the Enfield might not, in a modified manner, be made to do the work of the big guns."[128]

---

[127] Walker, 143
[128] Ibid, 144

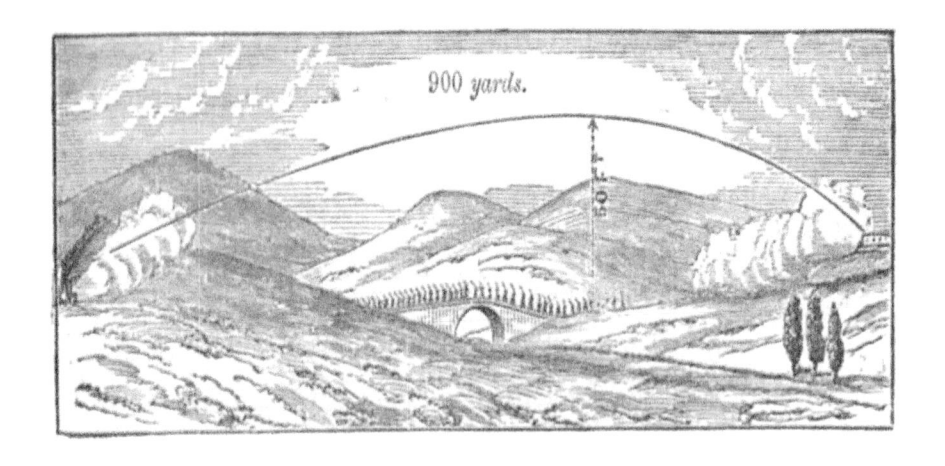

In his hypothetical scenario, Walker assigned 1200 out of his 1800 soldiers as his base of fire element, with the remaining 600 to constitute his assault element. The 1200 soldiers posted as the support by fire element open fire on the enemy work at the range of 900 yards, while the 600-man assault group continues their advance towards the bridge. "Firing volley after volley into the open area of this field work," Walker describes, "over the heads of their comrades, at the easy rate of two volleys per minute, thus 1,200 men would, in ten minutes, project no less than 24,000 Enfield bullets into the Redan." The rate of fire is reduced back to a leisurely two rounds per minute, but the volume of fire against the enemy position remains the same as Wilford's, at 40 rounds per second. As Walker describes the situation, the enemy would be unable to stand on the parapet or return effective fire while receiving such fire, and the assaulting element is able to cross the bridge in safety. "Beneath a hailstorm of lead it would be impossible for human life to exist." Walker was certain the enemy would either be forced to abandon the position and withdraw, or be annihilated.[129]

This is not to say that this sort of fire and maneuver envisioned by Wilford and Walker would have been effective or useful against near-peer forces on the battlefield; it is sufficient for our

---

[129] Ibid., 145

purposes to simply acknowledge that the concept had been borne out of having trained riflemen equipped with modern weapons capable of sustained, accurate fire. There is no concern for the "waste" of ammunition from "shooting at nothing." Only a few years earlier, such firing at a general area inhabited by the enemy would have been universally decried as an inexcusable waste of ammunition, but after Colonel Wilford, it is seriously promoted as a viable use of infantry fire by instructors at the School of Musketry. It is important to consider that nobody, least of all Walker or Wilford, were advocating such fire and maneuver tactics as the *primary* means of engaging the enemy. Instead, they were for use only in "extraordinary" situations, such as when artillery was not available to provide support. Even so, infantry's new-found ability to provide fire suppression to cover another infantry force moving forward in the attack was realized. These are the primitive, embryonic building-blocks of modern light infantry fire and maneuver tactics, and they emerged in the energetic minds of men like Colonel Wilford, who thought through the greater tactical ramifications of arming infantry with a modern weapon: the rifle-musket.

8
THE RIFLE IN THE INDIAN REBELLION, 1857

The rifle-musket was the first modern infantry weapon, but it was misused by untrained soldiers far more than it was employed by trained soldiers to anywhere near the extent of its theoretical capabilities. This is the inevitable consequence of taking the smoothbore musket out of the hands of a soldier, giving him a rifle-musket, and in most wars of the rifle-musket era, immediately sending him into battle with virtually no training. Modern historians blame the intrinsic faults of the rifle-musket, such as its slow rate of fire and great trajectory arc, for failing to launch warfare out of the Napoleonic into the modern era. But is it fair to evaluate the performance of the rifle-musket based on the battles and wars in which it was carried by, for lack of a better term, untrained amateurs? We have already seen the rifle-musket used effectively by trained soldiers, at long ranges and with decisive impact, in the Crimean War. Before we consider the rifle-musket in the Italian wars of 1859-1861 and the American Civil War, it is illustrative to consider another major campaign in which the rifle-musket was used by reasonably well trained-soldiers: the Indian Rebellion of 1857.

Ironically, the Enfield rifle itself (or more accurately, the ammunition for it) played a significant and controversial role in the outbreak of the Indian Rebellion. Enfield cartridges required a lubricating composition to soften the powder residue from the previous shot, and permit easy ramming of the next bullet. Before 1859 this lubricant was a mixture of beeswax and tallow, and the Muslim and Hindu sepoys of the Company's army were concerned that the tallow might be derived from religiously unclean pork or beef. Leaders of the rebellion claimed that the beef or pork tallow was deliberately applied to the cartridges by the British, in a perfidious scheme to defile the sepoys and make them break caste. It is not within the scope of this work to consider the causes and background of the rebellion, except that the Indian units refused the new weapons with their suspicious cartridges, and went into battle with Brown Bess. As several British writers noted, suppressing the rebellion might have been considerably harder if the rebels had waited to be completely armed with the Enfield. Instead, comparatively small numbers of European infantry (usually armed with the Enfield) and loyal native troops (usually Gurkhas) were able to defeat far larger rebel forces, often in strong prepared artillery-enhanced defensive positions. By 1857, many Hythe-trained officers from both the Queen's regiments as well as the Company had implemented musketry instruction in India. In general, these units were not thoroughly trained, but they had received some degree of instruction and were, overall, proficient in the use of the rifle. Although British authorities like Colonel Wilford would have considered them only marginally trained, they were certainly far better practiced with their rifles than virtually all American Civil War units.

In July 1857, General Sir Henry Havelock led the first significant relief column into Cawnpore with the objective of lifting the siege. His 1900-man force consisted of 1400 European troops from the Queen's 64th, 78th, and 84th Regiments of Foot, and the Company's 1st Madras Fusiliers.[130] Only the flank companies of the Queen's 64th and 78th had been armed with the Enfield, but the entire regiment of the Company's 1st Madras Fusiliers, made up entirely of about 400 European soldiers, had received the Enfield in May. Combined, about 600 soldiers of Havelock's column had the Enfield rifle, and the remainder were still armed with the older smoothbore. Just how extensively these soldiers had trained with their Enfields is unclear, but there were officers appointed in the Queen's regiments as musketry instructors. It is reasonable to assume that the rifle-armed soldiers in Havelock's army had received some measure of quality musketry instruction due to the effectiveness of their fire on the

---

[130] *The History of the Indian Revolt and of the Expeditions to Persia, China, and Japan, 1856-7-8* (W&R Chambers, London: 1859), 250

battlefield, far exceeding that of untrained soldiers. This intrepid little column advanced from Allahabad on 4 July, marching directly towards Cawnpore into the heart of the rebellion where vastly overwhelming numbers of sepoy rebels awaited them.

At Fatehpur, the rebels prepared a strong defensive position along the famous Grand Trunk-Road, the only practicable approach to and through the town. The fields on either side of the road were flooded with several feet of water. Here a force of 3,500 sepoys with 12 guns encountered Havelock's column, whose only avenue of approach was along the road, where the rebel guns and smoothbore musketry of their massed infantry could be concentrated.[131] Although his troops were exhausted from days of forced marching, the sepoys were sending 24-pound cannonballs into Havelock's bivouac (one British account bemoaned one of these balls that "smashed one of the camp kettles of the Sixty-Fourth").[132] The exhausted little army had to jump up from their rest and reform while Havelock made his plans for advance.

Of Havelock's subsequent attack, we have only British sources to rely on, and these sources in turn take nearly all of their details from the official reports of Havelock and his officers. These reports present the action at Fatehpur (and other battles) in the typical Victorian fashion, heaping glory upon themselves and feats of British arms for victories achieved over the most contemptible of enemies. We can be reasonably certain, however, of his general deployments. Havelock's artillery (only a few 9-pounders, under Captain Maude) was placed in the center with 100 Enfield riflemen from the 64th Foot. The Madras Fusiliers were deployed in open order as skirmishers, while the rest of the force advanced through the flooded plains in column.[133] Once

---

[131] Raikes, Thomas, *Services of the 102nd Regiment of Foot (Royal Madras Fusiliers), from 1842 to the Present Time* (Smith, Elder & Co, London: 1867), 27. William Tate Groom estimates the sepoy force at 2500.

[132] Forbes, Archibald, *Havelock* (London: MacMillan & Co, 1890), 115

[133] *Cassell's Illustrated History of England, Volume VIII* (London: Cassell, Petter, and Galpin, 1865), 442

formed, Havelock attacked.

Nearly every account, along with Havelock's own report, credits the Enfield rifle for securing a nearly effortless victory at Fatehpur. Captain Maude, who did not need to fear the smoothbore muskets of the sepoy infantry, audaciously advanced his artillery alongside the 100 riflemen from the 64th. The rest of the smoothbore-armed British infantry "pressed forward in support through the swamps and inundation, their advance covered by the Enfield skirmishers" of the Madras Fusiliers.[134] This was the first engagement between sepoy forces and British troops with the new Enfield rifle, and by all accounts the sepoys were shocked to be on the receiving end of effective musketry from long ranges. "The enemy saw a few riflemen approach," one account described, "but they knew little of the Enfield rifle and were panic-stricken with the length and accuracy of its range; they shrank back in astonishment."[135] Fire from the Enfield rifles "struck the rebels at ranges which filled them with amazement."[136] Under such effective fire, the sepoys began to break and were unable, or unwilling, to maintain their positions; their own return fire, with smoothbore muskets, was hopelessly ineffective at these ranges. Covered and supported by Enfield rifle fire, Captain Maude brought his 9-pounders up to the enemy's flank, "within two hundred yards of the hostile infantry."[137] In a remarkable early instance of combined arms in the assault, he sprayed the sepoy positions with grape alongside the Enfield riflemen of the 64th Foot. This was too much, and the rebels broke and retreated through the city, leaving their guns.

The day was not yet over. Shaken and astonished, the sepoys managed to rally and take up new positions about a mile behind Fatehpur. Havelock's troops were absolutely exhausted, but they advanced again with the same plan as before. The Madras

---

[134] Forbes, 117.
[135] *The History of the Indian Revolt,* 250
[136] *Cassell's Illustrated History of England,* 443
[137] Forbes, 117

Fusiliers went forward in open order, firing at the rebels from beyond the range of the Brown Bess carried by the sepoys. "Here the enemy appeared inclined to make a stand," wrote General Havelock's biographer, "but their hearts failed them when the Enfield bullets began to patter."[138] In a letter home to his wife Helen Groom, Lieutenant William Tate Groom of the Madras Fusiliers described the desperate final charge of "large bodies of cavalry" at an "impossible distance," attempting to move around the British flank. The Madras Fusiliers, in skirmish order, opened on the distant formations. "It was short, sharp, and very decisive affair. Our rifles completely upset their cavalry."[139] What remained of the sepoy cavalry turned and fled. Abandoning twelve guns, the rest of the sepoys broke and retreated, leaving the British in command of Fatehpur and the field. General Havelock, for his own part, attributed the nearly bloodless victory to "Almighty God" and, more temporally, to the Pattern 1853 Enfield rifle that enabled his troops to engage enemy beyond the range of their smoothbore muskets. The rifle and cannon fire alone, at long range, had driven the sepoys from Fatehpur; there was no need for a close-quarters bayonet fight. "Our fight was fought, neither with musket nor bayonet nor sabre, but with Enfield rifles and cannon: so we lost no men," Havelock wrote in his official report. "The enemy's fire scarcely touched us; ours, for four hours, allowed him no repose."[140] In a general order to his little army, Havelock asked "To what is this astonishing effect to be attributed?" He concluded directly: "To the power of the Enfield rifle in British hands; to British pluck, that good quality that has survived the revolution of the hour; and to the blessing of Almighty God on a most righteous cause."[141] The town of

---

[138] Ibid., 118

[139] Groom, William Tate, *With Havelock from Allahbad to Lucknow* (London: Marston & Co, 1894), 28

[140] Marshman, John, *Memoirs of Major-General Sir Henry Havelock, K.C.B.* (London: Longmans, 1876), 293

[141] *The History of the Indian Revolt*, 250. The battle at Fatehpur is often described as a "bloodless" British victory, as enemy fire inflicted no casualties to

Fatehpur was sacked, ostensibly for being sympathetic to the rebels.

General Havelock continued his advance to Cawnpore, fighting near-daily battles that increased in intensity and resistance. At each occasion, the Enfield rifles of the flank companies of the Queen's regiments and especially the Madras Fusiliers were the decisive element. Three days after Fatehpur, on July 15, Havelock's army reached the swollen Pandu Nadi river in full flood stage. On the far end of a "fine stone bridge" the rebels had established their strongest defensive position yet, with heavy guns and infantry placed to smother the choke-point with grape and musketry.[142] Havelock, marching light and fast, had no engineers or pontoons; the bridge was the only way forward. The overall situation is remarkably similar to Colonel Wilford's hypothetical scenario that required an attack across a bridge covered by an enemy defensive position.

On the Pandu Nadi, the sepoys "had for some days been diligently employed in intrenching themselves" near the bridge, and emplacing "a 24-pounder gun and a 24-pounder carronade, which swept the bridge and the great trunk road for 2000 yards beyond it."[143] The rebels were also preparing to blow up the bridge at the time Havelock's force arrived, leaving nothing to chance. It is worthwhile here to remember that the Indian regiments that joined the rebellion were professional soldiers of the East India Company, which at that time was in fact the largest standing army in the world. They received much the same drill and training as the regular British troops, and immediately before the outbreak of the rebellion, the sepoys were about to be issued the same P1853 rifle. In the rebellion, much of the discipline and unit integrity of the sepoy regiments were retained. Havelock was

---

Havelock's European troops (although twelve men were killed by sunstroke). Several killed and wounded from among the native troops of Havelock's column warranted no particular mention in his report, or in popular British accounts of the action.

[142] Forbes, 121
[143] Marshman, 299

not fighting clueless militia or ignorant new recruits, but rather the very same East India Company regiments that had been painstakingly trained, equipped, and armed by the British themselves. Their muskets, however, were smoothbores like those of the Russians at Alma, Inkerman, and Sevastopol. Sir John William Kaye, in his well-known and massive *History of the Sepoy War in India,* described the situation aptly: "In truth the enemy were terribly out matched. With all their gallantry in doing and their fortitude in enduring, what could 'Brown Bess' and the old station-gun do against our batteries and our rifles?"[144]

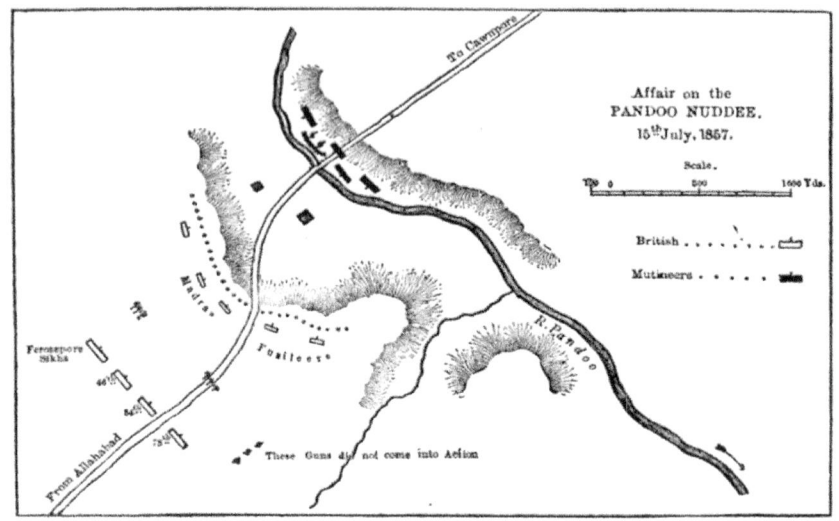

PLAN OF THE AFFAIR AT PANDOO NUDDEE.

General Havelock was an avid student of military history, and as a young lieutenant in the Rifle Brigade, he studied history and military science with Sir Harry Smith. With such interests, and prior service in the Rifle Brigade, Havelock must surely have been familiar with the achievements of Lieutenant Godfrey of the Rifle Brigade in the Crimea. By 1857, the routine of rifle-armed British infantry picking off artillerymen at long ranges had been accepted as a matter of course. The sepoy gunners, behind their big guns, expected the canalized and exposed British to

---

[144] Kaye, John William, *A History of the Sepoy War in India, 1857-58, Volume 2* (London: Longmans, 1896), 641

charge across the bridge, and be brushed off by 24-pounder blasts of grape. Instead, Havelock sent forward the Madras Fusiliers in open formation on either side of the trunk road. In Forbes' biography of Havelock, "A detachment of Enfield riflemen of the Madras Fusiliers moved down each of the lateral ravines, lined the banks, and opened a hot fire on the enemy's gunners and their cavalry supports."[145] In the account described in Havelock's memoirs, the Madras Fusiliers "opened a galling fire, and picked off the enemy's gunners, among whom and a large body of horse supporting them, the bullets were seen with the glasses to fall like a shower of hail."[146] The sepoy gunners, like the Russian gunners at Balaclava two and a half years before, were cut down in short order. After the battle, rammer and sponge staffs of the sepoy 24-pounders were found shattered; one account credits British bullets with severing them, and another believes it more likely the sepoys shattered their own rammers before spiking and abandoning the guns. Either way, the Enfield fire from the Madras Fusiliers all but silenced the guns. Sensing a decisive moment, without waiting for the main body to advance over the bridge, the Madras Fusiliers themselves charged across and captured the guns amid the piles of carnage.

The column proceeded to Cawnpore, fighting another major battle on July 16th and capturing the city. Again, Havelock used his Enfield rifles to great effect, silencing more artillery and defeating hapless enemy formations at long range. Lieutenant Groom wrote again to his wife after the Battle of Cawnpore, proudly exclaiming, "we showed them how to drive *whole regiments* out of snug positions by a couple of extended companies with Enfield rifles."[147] Another British officer at Cawnpore was even more succinct: "Our Enfield rifles did it all."[148] Havelock's march, with the frequent battles that dot the route of his advance, demonstrates the use of the Enfield rifle-musket by trained

---

[145] Forbes, 122
[146] Marshman, 299
[147] Groom, 35
[148] *History of the Indian Revolt*, 402

soldiers. In several instances, the sepoy positions were cleared out by effective long range rifle fire. Artillery was also silenced at long range, just as many theoretical strategists had predicted.

After taking Cawnpore, Havelock's column fought its way to Lucknow (using the Enfield to drive through superior numbers of enemy defenders along the way) in the first attempt to lift the siege of the British Residency there. Upon arriving at Lucknow, Havelock's small army was exhausted and depleted; the large numbers of sick and wounded at Lucknow forced Havelock to remain. The first "relief" of Lucknow merely became a reinforcement, and they would be forced to hold out until a much larger relief army arrived under Sir Colin Campbell (the hero of Balaclava). Yet Havelock's march on Cawnpore is the first significant example of the many large engagements of the Indian Rebellion where trained British and Company soldiers effectively used the Enfield rifle far beyond the capabilities of the older smoothbore musket. General Eyre and General Franks, in their similar campaigns, praised the Enfield rifle and its applications at long distances. By February 1858, when General Franks was using "the rank and file of an ordinary battalion armed with the new [Enfield] musket" to silence sepoy batteries, this feat had become rather old hat. Even so, the tactical revolution represented by the rifle-musket on these battlefields was clearly observed.[149]

William Forbes-Mitchell, a sergeant of the 93rd Highlanders (of Thin Red Line fame), recalled the second Lucknow relief column that fought savagely through strong resistance to reach the Residency. Originally dispatched from England for service in China, the 93rd was diverted to India at the height of the Mutiny. Out of all the British and Company forces that assembled under Sir Colin Campbell at Allahabad in October 1857, "the only complete regiment was the Ninety-Third Highlanders, over a thousand men, in splendid condition, armed with the Enfield rifle, and what was of more importance, well drilled to the use of

---

[149] Thackeray, xviii

it."[150] Three years had passed since the 93rd had stopped the charge of Russian cavalry at Balaclava; although the regiment had been much augmented by new recruits and transfers from other units, there were still many Crimean veterans in the ranks, including Sergeant Forbes-Mitchell's company commander, Captain Earnest Dawson. It is hard to imagine that, in October of 1857, there was any other regiment in the British Army (or any army) with more faith and confidence in the new rifle and its capabilities. The relief force reached Lucknow on November 14th and Campbell's army began fighting its way through a series of fortified positions towards the besieged Residency.

Most of this fighting from the 14th to the 17th of November was intense hand-to-hand combat as Campbell approached the Residency. On the morning of the 16th, the 93rd helped reduce the Secunder Begh, a large walled garden defended by two

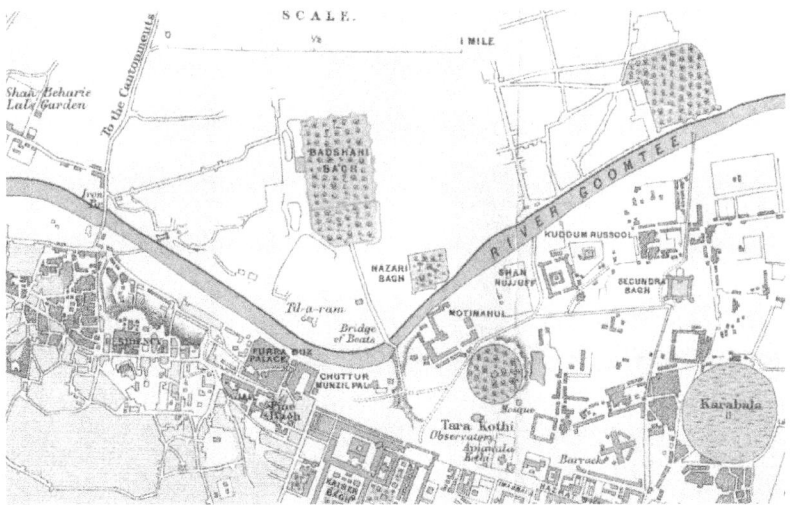

thousand sepoys. Campbell advanced that afternoon to the Shah Najaf, a multi-story mosque that contained the tomb of Ghazi-ud-Din Haidar Shah, the first king of Oude. After several assaults, a practicable breech was found by the 93rd and the Shah Najaf was taken. As night fell, the exhausted British slept soundly inside

---

[150] Forbes-Mitchell, William, *Reminisces of the Great Mutiny 1857-59, Including the Relief, Siege, and Capture of Lucknow, and the Campaigns in Rohilcund and Oude* (London: MacMillan and Co., 1893), 10

its thick, heavy walls. The outer perimeter of the Residency was only 450 yards away.

During the night, an alarming discovery was made: the sepoys had used the Shah Najaf as a powder magazine, and tons of gunpowder, had been heaped (literally) in the tomb. Forbes-Mitchell described "a big black heap about four or five feet high" of loose gunpowder in the central domed tomb, which was lined with dozens of barrels full of powder. The British soldiers at the Shah Najaf, under the direction of Captain Dawson, began a frantic effort to remove the five thousand pounds of loose gunpowder plus the barrels from their position. One spark is all it would take to turn the Shah Najaf into a smoldering crater. The precarious situation at once got a lot worse: once the first light of dawn on the 17th of November came, a sepoy battery across the Gomti River just outside of the walled royal gardens of the Badshahibagh opened fire on the Shah Najaf. "The enemy commenced firing shell and red-hot round shot from their batteries in the Badshahibagh across the Gomtee," Forbes-Mitchell remembered in his 1893 *Reminisces of the Great Mutiny 1857-58,* "aimed straight for the door of the tomb facing the river, showing they believed the powder was still there, and that they hoped they might manage to blow us all up."

The sepoy gunners at the Badshahibagh lobbing glowing red-hot shot towards the Shah Najaf were probably flush with excitement in anticipation of watching the tomb (and thousands of British soldiers of the relief column) vanish in a flash. To improve their accuracy, the sepoys moved their guns out of the Badshahibagh and advanced them closer to the river, to have a better chance of sending a round through the door of the powder-magazine tomb. Like the Russian battery firing upon the British 4th Division at Balaclava, the Indian gunners believed they had absolutely nothing to fear from infantry a thousand yards away. It is another of the great ironies of military history that these hapless sepoys were firing upon the 93rd Highlanders, with veterans of Balaclava and Inkerman still in the ranks, trained to the peak of efficiency with their Pattern 1853 Enfield rifles.

Forbes-Mitchell gives a concise account of what happened next:

> As soon as our rifles were cleaned, a number of the best shots in the company were selected to try and silence the fire from the battery in the Badshahibagh across the river, which was annoying us by endeavouring to pitch hot shot and shell into the tomb, and to shorten the distance they had brought their guns outside the gate on to the open ground. They evidently as yet did not understand the range of the Enfield rifle, as they now came within about a thousand to twelve hundred yards of the wall of the Shah Nujeef next the river. Some twenty of the best shots in the company, with carefully cleaned and loaded rifles, watched till they saw a good number of the enemy near their guns, then, raising sights to the full height and carefully aiming high, they fired a volley by word of command slowly given – *one, two, fire!* and about half a dozen of the enemy were knocked over. They at once withdrew their guns inside the Badshahibagh and shut the gate and did not molest us any more.[151]

Trained riflemen with the Enfield rifle-musket saved the Shah Najaf from being blown to bits. The Shah Najaf still exists in Lucknow today, as does the Badshahibagh, which is now the site of the University of Lucknow. Using satellite imagery, the distance from the Shah Najaf to the nearest walls of the Badshaibagh can be precisely measured as 1150 yards. This confirms that Forbes-Mitchell is entirely reasonable in his range estimation of "a thousand to twelve hundred yards" and also reveals something about the skill of the 93rd riflemen at judging distance.

---

[151] Forbes-Mitchell, 96

After silencing the guns outside the Badshahibagh, the 93rd and the other British troops at the Shah Najaf were given no respite. A large force of "at least six or seven hundred" sepoy infantry "made up their minds to retake the Shah Nujeef. They debouched on the plain with a number of men carrying scaling-ladders." Once again, it is worth remembering that the sepoy infantry were not clueless or incompetent; these were trained soldiers, and a determined counterattack with shock columns against exhausted British defenders in a precarious position was a sound decision by the established Jominian principles of the day. Against defenders armed with smoothbore muskets, the assault probably would have succeeded. Instead, like the Russian columns at Inkerman, the sepoys bravely charged into the jaws of a paradigm shift. For the second time in as many hours, the Enfield rifles of the 93rd Highlanders saved their position:

> Captain Dawson, who had been steadily watching the advance of the enemy and carefully calculating their distance, just then called "Attention, five hundred yards, ready – *one, two, fire!*" when over eighty rifles rang out and almost as many of the enemy went down like ninepins on the plain. Their leader was in front, mounted on a finely-accoutred charger, and he and his horse were evidently both hit; he at once wheeled round and made for the Goomtee, but horse and man both fell before they got near the river. After the first volley every man loaded and fired independently, and the plain was soon strewn with dead and wounded.[152]

Campbell finally reached the Residency on November 18th, and they successfully evacuated the besieged garrison. Lucknow was left in rebel hands until spring, when Campbell returned to lay siege himself. While the main British force besieged Lucknow, a brigade-sized element under Colonel Thomas Seaton was left at Fatehgarh as a sort of rear-guard, to prevent a sepoy force

---

[152] Ibid., 99

attacking from behind. Initially, Seaton was disappointed to be left in the rear, without taking part in the glorious inevitable victory with Campbell at Lucknow. Near the end of March 1858, however, Seaton learned that the rebels were taking advantage of his static inactivity at Fatehgarh, and were raising a large force. The rajah of Mainpuri, Taj Singh, had risen against the British; Seaton feared that he was planning to cross the Ganges and "raise the country" in the Uttar Pradesh Doab between the Ganges and Yamuna rivers. Seaton audaciously decided to launch an attack on the central rallying point of the rebels at Kankar, 22 miles away from Fatehgarh. He commanded "slender force" of just one thousand infantry, three hundred cavalry and five light guns.

The majority of his infantry were the 600 men of the 82nd Regiment of Foot, an ordinary line regiment armed with the Pattern 1853 Enfield rifle. In 1854, the regiment was about to rotate to India but the orders were canceled at the outbreak of the Crimean War, and (to the horror of its officers) the 82nd Foot "was called upon to give volunteers to other corps more fortunate in being selected for active service. A large number of the finest men and best soldiers transferred their names to the rolls of other regiments," leaving the 82nd Foot in Britain as a drained "donor" regiment.[153] One lieutenant colonel, transferring to the 34th Foot bound for Crimea, was authorized to bring the two flank companies of the 82nd Foot with him! The flank companies were typically the best and most experienced in the regiment. The men remaining in the remnant 82nd Foot would have assumed the shameful reputation as shirkers, cowards, malingerers, troublemakers, and such an empty husk of a regiment would be very low on the list of priority assignments for a musketry instructor graduating from the School of Musketry at Hythe. This fact may explain why the 82nd Foot was not assigned a permanent officer as Instructor of Musketry (and would not have one until an ensign was assigned in September 1859). The class sizes at Hythe were small, and the officers graduating from

---

[153] Jarvis, Samuel Peters, *Historical Record of the Eighty-Second Regiment, or Prince of Wales's Volunteers* (London: W. O. Mitchell, 1866), 76

Hythe with their certificates as Instructors of Musketry were valuable assets and naturally assigned to the regiments going to see active service.

Finally, in summer of 1855, the 82nd Foot received orders for the Crimea, since the winter of 1854-55 and subsequent combat losses had depleted the British forces outside Sevastopol and they needed every man. Quickly filled up to fighting strength with new recruits, 82nd Foot arrived four days before the surrender of Sevastopol, on 8 September 1855; they did not see any action, and the frustrated regiment remained in the Crimea (still without a permanently assigned musketry instructor) until August 1856. They were home in England only a few months before being ordered to China. With the outbreak of the Indian Rebellion, they were diverted again, this time to India, and with the officer instructor of musketry position still vacant. By this time, the 82nd Foot likely would have been assigned an NCO instructor of musketry, but only officers were named in the Army List. The 82nd Foot joined Sir Colin Campbell in the second relief of Lucknow, fighting alongside the 93rd Highlanders. During Campbell's subsequent siege of Lucknow and then being hotly engaged in the defeat of the Gwalior rebels at the Battle of Cawnpore. Afterwards, the 82nd Foot had been detailed to Seaton's small force, and now formed the majority of his infantry.

Seaton slipped out of Fatehgarh before midnight on April 6th and, marching fast, arrived at Kankar at dawn.[154] Seaton described the enemy position as a strong one, and although there was much "hubbub and confusion" in the sepoy camp, they quickly recovered from the surprise. The sepoys opened fire with three artillery pieces, and Seaton drove his small force forward, attempting to rapidly close the range and deploy. Suddenly, professional sepoy cavalry (probably a lancer regiment from the Bengal Army) moved out of the rebel camp. In Seaton's words, "In a few minutes, out from behind the left village, came two

---

[154] Malleson, George Bruce, *History of the Indian Mutiny, 1857-1858, Volume 2* (London: W. H. Allen and Co, 1879), 500

splendid bodies of cavalry, the finest and best mounted I had yet seen. One of these advanced toward my right the other and by far the largest went off to my left."[155]

The sepoy cavalry maneuvering on Seaton's left took advantage of a river bed, "and as they followed its bend to our left, their intention to take us in flank was evident." Like so many others, the sepoy cavalry "believ[ed] themselves to be perfectly safe from our fire" at distances beyond the old smoothbore range. Unfortunately for the sepoy lancers, "their tall spears, with bright points, showed their position distinctly as they moved along at a distance of 700 yards." Seaton quickly gave the order for two companies of the 82nd Foot to deploy against the cavalry threat to his left. Both Seaton and the historian of the 82nd Foot (who was also present) describe the companies wheeling to the left, and deploying into open order.

> This was speedily done, and the men, loosening their files, commenced firing calmly and deliberately, the musketry instructor having given the number of yards to fire at. In a few minutes -- before three rounds had been completely fired -- there was terrible confusion amongst the spears; and as Colonel Hale's hearty voice was heard encouraging his men to fire steadily, with good and deliberate aim, out bolted the whole of the cavalry as hard as they could, and were followed by our horsemen, and by shouts of derision from the whole line. This was the first time I had seen the Enfield rifle used in the field, and I thought it the very king of weapons.[156]

The "musketry instructor" was probably a

---

[155] Seaton, Thomas, *From Cadet to Colonel, Vol 2* (London: Hurst and Hackett, 1866), 279
[156] Ibid., 280. Colonel Hale, the commander of the 82nd Foot, had been a major in the regiment when, in 1852, he was one of three officers from the 82nd selected to be present at the funeral of the Duke of Wellington.

noncommissioned officer, and of insufficient rank to have been named in the 1857 or 1858 Army Lists alongside the regiment's officers. It is also likely that the 82nd Foot would have been given instruction by the musketry instructors of other regiments, and the lack of a permanent officer instructor of musketry did not exempt the regiment from conducting the mandatory annual training and rifle qualification. At any rate, Seaton's account confirms that the regiment had a musketry instructor of some kind, and it follows that the regiment had also received some measure of musketry training. The humble 82nd Regiment of Foot demonstrates that even the most ordinary and neglected line units could (and did) effectively use the improved long range accuracy of the rifle-musket in combat. This capability was not limited to designated experts, or a highly-trained specialized light infantry unit: the British Army's system of instruction was proven, by these and countless other documented engagements during the Crimean War and Indian Rebellion, to extend the capabilities of the rifle-musket to the common line infantryman.

After cutting up the sepoy cavalry at Kankar, Seaton turned his line of fire toward the primary enemy position and routed the much larger force (probably between 3,000 to 5,000). One of Seaton's officers recalled that, "The [rebel] infantry were not in any order and amounted to several thousands. The Brigadier [Seaton] commenced firing at them from a long range, and our shot stirred them up and sent them further off. As we advanced, the 82nd let in with the Enfield, and a lot were knocked over."[157] Seaton's loss was three killed and 17 wounded; the Indian casualties were estimated at over 250 killed and at least as many wounded. Like William Howard Russell three years earlier, Seaton described the Pattern 1853 Enfield rifle as "the king of weapons" and credited the new rifle's capabilities for the easy victory.

Seaton, and the 82nd Foot, would fight another decisive engagement a few months later at Bunkagong, on October 8, 1858. It was one of the last field engagements of the Indian

---

[157] *Narrative of the Indian Revolt* (London: George Vickers, 1858), 444

Rebellion. Rebels had surrounded Powain and were burning nearby villages, and once again Seaton sortied early in the morning on a 12-mile night march. The timing of the march was perfect and, just like at Kankar, his small force arrived outside the enemy camp at Bunkagong just as first light was dawning. Unwilling to be caught by surprise, the rebels had posted a mounted picket and they sounded the alarm as Seaton's infantry, formed into line, began advancing into artillery fire. Shrapnel shells burst overhead, and Seaton described a near-miss from "a couple of shells, aimed beautifully, sending bullets rattling around our ears in fine style." With Seaton's infantry deployed into line, the rebels sent large forces of cavalry galloping around the British formation, in a doctrinally-sound attempt to turn the flank. Under the old Napoleonic battlefield paradigm, turning and flanking mere helpless infantry deployed in line (and not in square) was exactly what the tactical situation called for. "The infantry I formed into line behind the guns," Seaton wrote a few years later.

> The moment our artillery commenced firing, the enemy's cavalry moved forward on both flanks; and as soon as they got within 700 yards, I made the 60th and the 82nd try the power of their Enfield rifles on them. I was watching the cavalry on the left, for it was the largest body by far. They were coming round the end of the morass, to get into our rear by the road on our left. As soon as they got clear of some intervening trees, the light company of the 82nd began to fire on them, and we could see the men's heads and shoulders, and here and there a horse's head above the cultivation in the fields. The effect of this fire was curious. The impetuous horsemen suddenly pulled up and looked about, astonished and alarmed at the storm of bullets raining upon them they knew not whence, and hitting them with such force. The noise, confusion, and jumble in their ranks, horses rearing and stumbling,

and men falling, presented such a scene as is rarely witnessed, and in almost as short a time as I have taken to describe it, the whole mass turned and fled.[158]

It was another lopsided victory. The rebels lost 300 men, while Seaton lost two men killed and twelve wounded. Just as the Enfield rifle in trained hands had proven to be a combat multiplier in Havelock's army at the beginning of the Mutiny, the effective use of the rifle at Bunkagong ended the bloody and tragic military phase of the Rebellion.

These numerous accounts from the singularly brutal and intense battles of the Rebellion demonstrate that the British and Company troops, equipped with the rifle-musket and (more importantly) highly trained in its use, regularly engaged enemy forces at ranges beyond the old smoothbore. They also regularly judged the distance, and set their sights accordingly. The fire, even at long distance, was considered effective even if most of the bullets missed and only a fraction actually hit anything. Opposing infantry on the battlefield were cut down; infantry fixed defensive positions were driven out. Artillery was silenced on numerous instances, and the sepoy cavalry probably fared the worst of all, being inevitably shot to pieces in nearly every recorded encounter with British troops that had the Enfield. It is worth noting that these combats were rarely fought in "ideal" conditions, but instead in some of the most appalling and horrific circumstances imaginable, by exhausted soldiers who in many cases had been engaged in non-stop brutal fighting against a determined enemy for days or even weeks.

Thackeray, writing the introduction to the revised version of his *Soldier's Manual of Rifle Firing* in 1861, remarked that "since the first edition of this work was published, we have had in the Crimean and Indian wars many examples" of his prediction that the rifle-musket would change the battlefield forever. He emphasized that "the example should not be lost on our military or regimental authorities." As he wrote this, southern states on

---

[158] Ibid., 308.

the other side of the Atlantic were seceding from the Union, heralding the imminent start of the American Civil War.

9
MEANWHILE, IN FRANCE...

Before considering the rifle-musket in the American Civil War, we must return to France, the birthplace of the rifle-musket. and the epicenter of the Napoleonic bayonet-focused shock tactics that mesmerized most Western militaries for decades. It is a bit of historic irony that the French invented and adopted the Minié rifle with high expectations, only to find the rifle-musket less impressive in practice than in theory, and revert back to the familiar tactics of Napoleon. In the late 1840s, Claude-Étienne Minié invented his infamous missile that made a rifled infantry weapon practical. For the next several decades, French theorists would struggle with the matter of how best to employ the rifle, while simultaneously avoiding its greater destructive power.

In the 1850s, the British Army gradually adopted fire tactics and British tactical theorists envisioned battlefield success based on effective musketry at distances considerably beyond the old smoothbore range. By the late 1850s, British tactical thought had elevated the rifle to preeminence and keenly anticipated the combat multiplier of having an entire army trained at judging distance and long-range fire. The French, meanwhile, took a fundamentally different course that was skeptical of long range rifle fire and placed strong reliance on the decisive effect of the bayonet and maneuver on the battlefield. By the late 1850s, common British privates in ordinary line regiments were being issued Enfield rifles with 900-yard backsights, but the French line infantry were given a Minié rifle with a fixed block sight suitable only for short range work. Only specialized units received rifles with adjustable backsights. The common French soldier was not expected to fire beyond about 300 yards, and idealized French tactics emphasized units moving in rapid and athletic fashion, pressing with flexible movements to get close enough for the bayonet.

These tactics descended from the Napoleonic armies raised by conscription in the early industrial age. Large armies of citizen conscripts opened new strategic options and enabled Napoleon to readily smash the old, slow, armies of the rest of Europe. The column formation was well suited for Napoleon's conscripted infantry. Maneuver in the column was much simpler than the line, and the column was ideal for covering distances rapidly. Even Jomini somewhat famously wrote in the *Art of War* that "The French, particularly, have never been able to march steadily in deployed lines." A column quickly delivered large numbers of troops across a battlefield in an easily controlled and directed manner. While the column was vulnerable to fire, particularly roundshot from artillery and, to a lesser extent, the short range musketry of a line, the simplicity and utility of the column outweighed the vulnerabilities. It was accepted as a matter of course that driving a column into an enemy position would incur a significant number of casualties, but the Napoleonic *levée en masse* provided an almost inexhaustible number of replacements. Green soldiers, with minimal training, could be packed into a column; they would be unable to see or feel what was happening outside the dense formation, and were physically shielded from fire by the ranks and files on the exterior. Battalions of raw recruits could be maneuvered in these column formations across the carnage and shock of a battlefield, driven forward by their own mass.

The French attempted to distill all the experience of Napoleon in the updated 1831 *Ordonnance sur l'Exercice et les Manoeuvres de l'Infanterie,* the standard manual of infantry drill. The tactics were simplified and conservative; Paddy Griffith described the 1831 *Ordonnance* as "retrospective rather than innovative," and the French tactical system "did not in itself seem to change very much."[159] Some historians have considered the post-Napoleonic French Army in the forty years following Waterloo as waning, intellectually vapid institution, with some justification. Two significant events injected renewed debate into the French Army

---

[159] Griffith, Paddy, *Military Thought in the French Army 1815-1851* (Manchester, UK: Manchester University Press, 1989), 125.

and *forced* a reconsideration of strictly Napoleonic tactics developed for European battle. The first was a long and particularly challenging war in Algeria, and the second was the unexpected appearance of the long-range rifle-musket.

During the French conquest of Algeria 1830-1847, the French infantry encountered an enemy vastly unlike the enemies of Napoleon a generation earlier. Smoothbore muskets, hammering away at elusive, shadowy opponents, accomplished nothing. The Algerian resistance refused, for the most part, to offer conventional battles in the field, and resorted to mobile guerrilla tactics. Thomas Robert Bugeaud, who had learned to respect the fire of British line infantry in the Peninsular Campaign, developed the new tactics and strategies for chasing down and defeating clever, fast-moving, and tenacious Algerian fighters. The formative experiences of the Algerian war would leave a long mark on the French military identity: the Zouaves, *le pantalon rouge,* and an appreciation for the psychological domain of war, were products of the grueling Algerian campaign. The utility of a brand-new weapon, the rifle-musket, was appreciated very early on, and the first rifle-muskets that operated on the Delvigne chamber principle were employed in Algeria in the late 1830s. Elite troops such as the *Chasseurs à pied* were able to use rifle-muskets to increasingly great effect against the Algerian guerillas.

The lessons the French Army took from Algeria tended to confirm the need for two types of infantry: the ordinary infantry masses, and a highly trained light infantry. With the rifle-musket came the new threat of long-range fire; there was furious debate over how to address the modern battlefield and its unfamiliar threats. "Senior officers," Griffith wrote, "were often heartily sick of young subalterns who styled themselves as innovators — particularly those who pressed for 'light infantry' skills such as shooting, fieldcraft, gymnastics," etc.[160] By and large, the French chose to leave the whole business to the *Chasseurs à pied.* In many ways the *Chasseurs* were the embodiment of the French military character. In the French service academies at St. Cyr, Vincennes,

---

[160] Ibid.,126.

and elsewhere, the varied disciplines of gymnastics, fencing, bayonet drill, and rifle shooting were "prized as national military qualities in the same way as the English might value their 'solidity.'"[161] In the early 1850s, the *Chasseurs* received the new Minié rifle. They were extensively trained in musketry, and received rifles with long-range sights calibrated for shooting out to 1000 yards. The ordinary French infantry, on the other hand, were issued rifles with a single block sight calibrated for about 200 yards; they were not expected to fire significantly beyond that distance, and were trained to close the distance rapidly to engage with the bayonet. By the mid to late 1850s, long range rifle shooting at anything over 200 yards was strictly the province of the elite light infantry units. Armand-Jacques Leroy de Saint-Arnaud, the French Minister of War, veteran of the Algerian war, and the commander of the French troops during the opening stages of the Crimean War, sought to limit the preliminary skirmish firing conducted by the infantry prior to the bayonet assault. The new French emperor, Napoleon III, wanted even less by 1859.[162]

Two years earlier, in 1857, Napoleon III established a military camp at Châlons. It had a practical purpose as a training-ground for French troops, but served a duel propaganda role to show off the new Imperial Army before French citizens and foreign observers, press, and tourists. Châlons remains a major French military base to this day. In 1859, as war-scare tensions between France and Britain were nearing their zenith, an anonymous British writer for *Blackwoods Edinburgh Magazine* visited Châlons and wrote a comprehensive description of what he saw there. To any observer, but particularly to a civilian, the choreographed gymnastics, the rapid movements, the agile sprinting assaults, and smooth practice with the bayonet would be extremely impressive. The *Blackwoods* writer was astonished by the French drill, and urged the British Army to follow the French example and copy their quick, athletic attack columns. "Mobility," he wrote, "would

---

[161] Ibid., 108
[162] Ibid., 128.

seem most consistent with the war inventions of the present time – most adapted for the tactics demanded by the long range of projectiles." The French rifles, he noted, "had only one sight, fixed so as to give a convenient point-blank range," except for "the *Chasseurs* alone, who have moveable sights capable of being adjusted for long distances." All aspects of the French military method were built around rapid movement, agility, and shock:

> The lessons of the desert were adopted as laws, and the Crimea was the harvest-field of seed sown in African marches and combats. It is with them a creed, that rapidity of movement and velocity of attack are more important elements in tactics than solidity, if it be accompanied by immobility. That quickness and locomotion are more valuable qualities in the soldier than compactness or steadfastness.[163]

    The persistence of these tactics, brought to a culmination in the dashing *Chasseurs*, owe much to the most influential military theorist of the rifle-musket period, the Swiss-born Baron Antoine-Henri Jomini. As *the* great strategist and theorist in Europe, the comprehensive codification of Napoleon Bonaparte's campaigns in Jomini's *Art of War* was popular reading among the civilian-turned-soldier officer corps of the American Civil War. General J. D. Hittle observed that "many a Civil War general went into battle with a sword in one hand and Jomini's *Summary of the Art of War* in the other."

    Jomini purportedly told Napoleon III (who Jomini humbly identified only as a "distinguished person") in 1851 that the rifle "might, probably, exert a certain influence on the details of tactics," but that "victory will always be secured by the same principles which gave it to the great captains of all ages -- Alexander, Caesar, Frederick the Great, and Napoleon."[164] Jomini is best known – then and now – for his thoughts on

---

[163] Blackwoods Edinburgh Magazine, "Chalons.—The Camp," 272
[164] *Colburn's United Service Magazine,* Part III, 1864, 11

strategy and grand tactics, which were the first two of his six "great principles" of war. He did write extensively on tactics, however, and remained absolutely committed to the columnar tactics of Napoleon long after the rifle-musket had been introduced in the armies of Europe and the Americas. He was extremely skeptical of fire tactics. In his chapter on tactics, Jomini gave all the weight of his authority to the statement, "I cannot imagine a better method of forming infantry for the attack than in columns of battalions."[165]

To Jomini, the ideal formation of troops for battle would have the characteristics of solidity, mobility, and momentum. The small column met all of these criteria, and Jomini argued strenuously against the use of large, massive columns because of their extreme vulnerability to fire. He observed that under the "cool and deadly" fire of the British, "French columns did not succeed so well at Talavera, Busaco, Fuentes-de-Onore, Albuera, and Waterloo." The destruction of the French by British musketry at these battles did not convert Jomini to *Feuertaktik*. Instead, he concluded it was because the French had employed the wrong *kind* of columns. He continued:

> We must not, however, necessarily conclude from these facts that the advantage is entirely in favor of the shallow formation and firing; for when the French formed their infantry in these dense masses, it is not at all wonderful that the deployed and marching battalions of which they were composed, assailed on all sides by a deadly fire, should have been repulsed.[166]

Waterloo, in particular, was cited by the defenders of fire tactics as evidence of the supremacy of fire over shock. Jomini strongly disagreed. While Wellington's fire of infantry in two ranks with "murderous effect" at Waterloo was remarkable and

---

[165] Jomini, Antoine-Henri, *The Art of War,* trans. G.H. Mendell and W. P. Craighill (Philadelphia: J.B. Lippincott, 1862), 299.
[166] Ibid., 298.

"created doubt in some minds as to the propriety of the use of small columns," Jomini insisted that the French lost because they failed to use his mixed order. Wellington defeated massive, dense columns, not Jominian small columns. The Swiss strategist even had the chance to speak with the Duke of Wellington in 1823 at Vienna, and Jomini tried to convince the Iron Duke that Napoleon would have won at Waterloo if Jominian small columns had been used. Ultimately, Jomini concluded that the French defeat at Waterloo "was due neither to the musketry fire nor to the use of deployed lines by the English." Instead, it was due to "accidental causes" such as deep mud that retarded French advances, the deep columns used, the uncoordinated and unsupported infantry and cavalry attacks, and (last but not least) the unexpected arrival of the whole Prussian army at the decisive moment of battle.

Small columns of limited depth, moving rapidly, yet still delivering sufficient mass to take an objective, was Jomini's ideal formation. He advocated the "mixed order" of troops with a firing line of skirmishers in front of stacks of small columns and pointed to Napoleon's use of similar formations, although the Emperor did not, in fact, seem to use the famous *l'ordre mixte* very often in the way Jomini described. Even in Jomini's preferred mixed order attack, only the first line of skirmishers in front were permitted to use their muskets. He reminded commanders that "a column of attack is not intended to fire, and that its fire should be reserved until the last," and that if troops "begin to fire while marching, the whole impulsive effect of its forward movement is lost." It was not really necessary to fire at all in some cases, according to *The Art of War*. Jomini recounted that "positions have often been carried by Russian, French, and Prussian columns with their arms at the shoulder and without firing a shot," and credited this as "a triumph of *momentum* and the moral effect it produces." In his advice to generals and commanders, Jomini said succinctly that "small columns have always succeeded wherever I have seen them tried."[167]

---

[167] Ibid., 294-298.

Jomini's readers would, if nothing else, come away from *The Art of War* with a conviction that *Feuertaktik* was absolutely the least advantageous mode of combat, and the worst possible form of attack. Only in passing, after exhaustively listing various forms of column attacks, does Jomini mention employing firepower. "Finally," he writes with almost tangible reluctance, "a last method is that of advancing altogether in deployed lines, depending on the superiority of fire alone, until one or the other party takes to its heels – a case not likely to happen." In the 1850s, Jomini added an appendix to *The Art of War* and addressed the adoption of rifle-muskets by most European armies. His convictions were fundamentally unchanged, except that he emphasized even smaller, faster columns and even less pausing to fire. Simply put, fire tactics would never win battles: "in spite of the improvements in firearms," Jomini wrote, "two armies will not pass the day firing at each other from a distance; it will always be necessary for one of them to advance to the attack of the other." He assured commanders that "victory may with much certainty be expected" by the army whose general "possesses the talent of taking troops into action in good order and boldly attacking the enemy."

Jomini valued the brilliance of the commander and the ability of a general to see through the complexities of a situation and apply the artful principles of war to his advantage. By the proper understanding of strategic geometry, a successful commander would achieve victory through maneuver, converging his forces upon the enemy at the most decisive time and place; the actual fighting of the battle, which would be won thanks to superior numbers in advantageous positions, was simply the last piece of the process. The commander would also select the tactical formations best suited for the battle. Ideally, the enemy should be beaten before the first shot is fired, and the battle merely confirms that fact. Except for skirmishers and elite light infantry, Jominian tactics had no need for trained riflemen in the ranks. The individual training of the line soldier was focused primarily on marching formations through interminable evolutions of drill, so that the soldier (acting as part of a larger mass) would be able to perform a number of tactical movements. Jomini specified seven changes of tactical formations that, at a minimum, troops should be proficient in, such as "marching in deployed battalions arranged checkerwise." It was also important that troops be "solid" in battle, with a collective *morale* that, by itself, might overpower less-cohesive enemy formations. After the adoption of the rifle-musket by most European powers, Jomini did not foresee any dramatic changes in strategy or tactics. He famously decided that "the improvements in fire arms will not introduce any important change in the manner of taking troops battle."[168]

The shock tactics of Jomini and the elite skills of the *Chasseurs* finally had the opportunity to be tested on a large scale in 1859, in the Second War of Italian Independence. The French, and their Sardinian allies, engaged the Austrians on the plains of northern Italy during the summer of 1859. At the outbreak of war, on May 12, 1859, the Emperor Napoleon III addressed his soldiers, assuring them that "the new weapons of precision are only dangerous from a distance, and they will not prevent the bayonet from being, as in the past, the terrible weapon of the French infantry."[169] True to the doctrines of Jomini and the *Chasseurs,* these were the tactics employed by Napoleon III at Magenta and Solferino, the two major battles of the Italian War. In both of these battles, fought three weeks apart in June of 1859, the French tactics were characterized by spirited attacks with assault columns.

In the Austrian Army, small steps had been taken towards fire tactics. The Austrians had adopted a new rifle-musket, the Model 1854 Lorenz in .54-caliber, and Austrian infantry manuals assumed that the rank and file would have received basic marksmanship instruction. The M1854 Lorenz was issued in two models. One variant had a long range sight, to be issued to the light infantry units that would be trained to a higher level than the ordinary soldiers of the line. Unlike the British sights on the

---

[168] Ibid., 359.
[169] Adam, Charles, *La Guerre D'Italie* (Paris: N.-J. Philippart,1859), 180.

Enfield rifle, the Lorenz sights were cleverly set up so that a trained soldier could fire at ranges from 300 to 900 *Schritt* simply by adjusting the sight picture, thus saving time. The standard variant had a fixed block sight, calibrated to 300 *Schritt*, for the line infantry. When shooting at targets at closer range, it was expected that the soldier would have been trained to aim *lower*, to counteract the arcing trajectory of the bullet. It was a fine rifle and known to be extremely accurate, but unfortunately for the Austrians, the capabilities of the rifle were squandered on the polyglot, polyethnic Habsburg army. At Magenta and Solferino, most of the Austrian troops were green recruits or reservists, who had no experience with a modern rifle. Any training they had previously received was with the old smoothbore, and on the eve of battle many of the untrained Austrian troops were issued the new rifle "still slick with factory grease."[170] Moltke observed in his analysis of the war in 1863 that "a large proportion of the Austrian Army was recruits, and the serving soldiers did not know how to use the new rifle that was handed to them."[171] Another German history of the 1859 war was more specific:

> They [the Austrians] were not prepared for this war; the number of reservists in the regiments were not doubled, but quadrupled, and they were overwhelmed with recruits. A good part of the men had never taken a single practice shot at the target, before they were sent to fight; they were defenseless and only targets for the enemy. The steadily increasing losses of the Austrians compared to their opponents can be explained by the fact that the less capable the troops, the greater the casualties are. Judging distance was never practiced, so the fundamental basics of tactics

---

[170] Wawro, Geoffrey, *The Austro-Prussian War* (New York: Cambridge University Press, 1996), 11
[171] Moltke, Helmuth von, *Der italienische Feldzug des Jahres 1859* (Berlin: Mittler und Sohn, 1863), 95, 174

and capabilities of the weapons [*Waffenwirkung*] were completely neglected.[172]

Against the untrained and poorly lead Austrians, the French storm columns crashed into the Habsburg lines at Magenta and Solferino, and the French stood in possession of the fields after many hours of carnage. French light infantry, maneuvering quickly, famously closed the distance and fought with the bayonet. The phrase *élan vital* had not been coined in 1859, but the French tactics in the era of Solferino certainly placed great emphasis on the energy, speed, and decisive action of the soldier, with the bayonet as the crucial weapon of decision. Highly-trained French *Chasseurs* did fire from relatively long range on the hapless Austrian artillery at one point, although the use of long-range rifle fire in 1859 was the exception rather than the rule. Most of the fighting was at close range, and the battles were marked by an unusual amount of savage hand to hand combat. The Austrian artillery, composed entirely of smoothbore guns, was also hopelessly outclassed by new French rifled field artillery, which inflicted devastating losses.

These accounts reverberated through Europe (and even the United States), because Solferino was fought on an open plain, between the national armies of great powers, each led by their crowned emperors. For most European military theorists, the war of 1859 with its massive field-battles was the example to study. The Crimean War provided no such examples of decisive battles on the field, with armies maneuvering against each other in a grand scale, dripping in glory; the campaign at Sevastopol was studied primarily as a siege-work against a static, fortified place. The rebelling sepoy regiments of the East India Company were never considered to be the equal of soldiers in standing European armies. As a result, the decisive French victory in the campaign of Solferino against the standing army led by the crowned Emperor of Austria only heightened the great interest in French tactics and weapons, not least of which was a very

---

[172] *Der Krieg in Jahre 1859* (Bamberg, Germany: C. C. Buchner, 1894), 8

limited reliance upon the rifle-musket as a long-range projectile weapon. The fact that a large percentage of the armies at Solferino were still armed with the old smoothbore was not lost on the critical observers; it seemed quite apparent that the French could have won with or without rifle-muskets. At the Battle of Magenta, fought as part of the Solferino campaign, French troops assaulted and captured an Austrian artillery battery with the bayonet alone, never firing a single shot. The battles of 1859 were obsessively studied by military thinkers around the world, who had been desperate for just such a trial by combat of their various theories. Most concluded that the lessons of 1859 taught that while the rifle may have *helped* the French in some places, it was not decisive. The debate over which tactical system was superior -- the rifle or the bayonet -- would ebb and flow until the First World War. The impact of Solferino would endure in the French Army, with tragic consequences for the French troops of 1914 in bright red trousers and blue tunics, who expected the great moral force of their bayonet charge to overcome German machine guns.

The impact of Solferino reinforced a near religious devotion to the bayonet and shock tactics in France, and even convinced the Austrians to abandon their tepid experiment with rifles and fire tactics. Most of Europe, and the United States, fawned over the nimble *Chasseurs* who captured entire Austrian units without dirtying the barrels of their Minie rifles by firing them. Between 1859 and 1864, *only* the British Army persisted in their gradual embrace of fire tactics. General Hay, Colonel Wilford, and their disciples at the School of Musketry advanced a general theory of fire tactics that, for convenience, I will describe as the "Hythe School," summarizing the British Army's tactical thought from about 1856 to 1866. The Hythe School vigorously argued that the average soldier, with some training, could utilize a rifle-musket out to the full extent of its capabilities, at least to 600 yards. As impressive as the French bayonet-centric tactics were, the instructors at Hythe were confident that any French columns would be destroyed by fire long before their bayonets came close enough to do any harm to the battalions of British riflemen.

The Hythe School preferred to look to the Crimean War and the Indian Rebellion for examples of the rifle-musket's impact on the battlefield: the British quickly decided that "the capabilities of long-range musketry were not fully tested on the Italian plains," where besides "the majority of the troops on either side [at Solferino] carried weapons of the old fashion."[173] The British Army was not at all surprised by the rifle-musket's poor performance at Solferino and Magenta. The French used them sparingly, and the Austrians had received no training in marksmanship or judging distance. Indeed, men like General Hay and Colonel Wilford believed that Solferino proved the importance of the Hythe system of musketry instruction. Without a system of instruction, even the finest rifle would be useless in the hands of an untrained recruit, whether that recruit wore Austrian white or British red. It was a system that required a corps of highly trained instructors, a strictly implemented program of marksmanship training, an expensive qualification process that required every soldier in the army to expend 90 rounds a year, and the finances and infrastructure to provide such a quantity of ammunition. The British system made the institutional decision that the common soldier was intelligent and capable enough to judge distances, utilize adjustable sights, and function with some degree of independence on the rifle-dominated battlefield. Among the advocates of the Hythe School (like Hans Busk, Arthur Walker, Watson, Hardinge, etc.) there was great optimism that technological advancements would only make the rifle more effective in time, by lowering the trajectory and reducing the need for near-absolute precision in judging distance. With the hindsight of history, we understand today that this eventually did happen with the adoption of smokeless powder and high-velocity bullets. The bayonet did not surrender its revered place in infantry combat quietly or quickly; the bayonet assault course for U.S. Army soldiers in basic training was not discontinued until 2010.

---

[173] *Saturday Review*, Volume 9, January 14, 1860, 55

On the other hand, the French school of Napoleon III was simultaneously conservative and audaciously innovative. It took a sober, pragmatic, and realistic look at the battlefield, and concluded that in the confusion, noise, and terror of modern combat, soldiers would not be fiddling with the sights on their weapons or calmly estimating the distances to their enemies. The French, in this period, were the first to take a serious consideration of the psychological impact of battle, and how to train soldiers to fight (and fight well) in spite of the stresses and fear on the battlefield. If the infantry soldier had to be trained to a peak of efficiency on a weapon, then that weapon must be the bayonet. The New York-published *Military Gazette* of January, 1859 gives a fair summary of the Napoleon III School on this point, immediately before the Italian war:

> No matter how perfect the gun, men, in the heat and excitement of battle, will hardly be deliberate in aim, or effective enough in firing to stop a charge of determined men: the bayonet, with most of mankind, will always be the queen of weapons in pitched battle; only for skirmishing, for sharp-shooting, and artillery, will the rifle equal theoretical expectations.[174]

The French school believed the British overestimated the effectiveness of the rifle (and the ability of a soldier to use it). The favor was returned by the British, who conjectured that "the French, perhaps, have refined too much upon this part [bayonet drill] of the soldier's training."[175] The French infantry rifle of this period, with its nonadjustable block 200-yard fixed sight, reflects the philosophy that the influence of the rifle "might" be useful in some situations. In other words, it's good to have them around, even if they may not be needed. Jomini, for his part, offered no further insights on the rifle until after the American Civil War and the Austro-Prussian War of 1866, when breechloaders were

---

[174] *Military Gazette,* Volume 2, Number 1, New York, January 1859
[175] *Saturday Review,* Volume 9, London, January 14, 1860, 55

rapidly superseding the rifle-musket and the tactics of the smoothbore era were finally abandoned for good.

A year before the outbreak of the American Civil War, the enthusiastic yet untrained volunteers of Giuseppe Garibaldi's famous 1860 campaign of the Italian *Risorgimento* formed a remarkable parallel to future Civil War armies. Garibaldi's volunteers were armed with an assortment of weapons, but international supporters financed the purchase of P1853 rifles which armed a number of Garibaldi's units. They were exuberant and motivated, but except for the veterans of 1859 in the ranks, the volunteers had no military training to speak of. One example was the British Legion of volunteers under Colonel John Dunne; "Dunne's men, who had had a few weeks' drill, did not know how to use the sights of their Enfield rifles."[176] One of the English volunteers recalled later that "Dunne's men had no target practice" and had to be content with firing blanks only, "so they were not able to make the best use of their weapons." Ultimately, according to the volunteer, "it did not matter much; elsewhere the contest was decidedly at close quarters."[177] Captain Charles Stuart Forbes, R.N., one of the English volunteers, similarly described Garibaldi's force as generally "armed with Enfields, but few knew how to develop the use of these deadly weapons, the sights being deemed a superfluity."[178] J. A. Dolmage was another British Army officer who took leave from his regiment to join Garibaldi's volunteers; he would have certainly been familiar with the British Army's musketry instruction, in contrast to the lack of training of the volunteers. For this style of fighting, with untrained yet recklessly passionate volunteers, "the smooth-bore weapon was not at all an unsuitable fire-arm for the tactics of Garibaldi, which was to get as quickly as possible to close quarters

---

[176] Trevalyan, George Macaulay, *Garibaldi and the Making of Italy, June - November, 1860* (Longman's, Greene & Co, London: 1914), 82
[177] Ibid., 328
[178] Forbes, C.S., *The Campaign of Garibaldi in the Two Sicilies* (William Blackwood & Sons, London: 1861), 92

with the enemy, when the bayonet became of use." Dolmage went on to observe that "the rifle was not of service to the soldier until he had been properly trained to use it."[179] It was no surprise to anyone, least of all the professional British officers volunteering in Garibaldi's ranks, that these untrained soldiers, even though they were armed with P1853 Enfield rifle-muskets, were not able to revolutionize warfare with the new weapons.

The French school also found advocates on the other side of the Channel, at least to some extent. In 1857, Lt. Colonel Matthew Dixon, a Royal Artillery officer and head of the Government facility that made Enfield rifle-muskets, read his paper, *The Rifle -- Its Probable Influence on Modern Warfare*, at the United Service Institution. Even though the Enfield rifle had been adopted as the universal infantry weapon for the British Army some years earlier, as a thoroughly conservative Victorian, Dixon had strong reservations about the capacity of common soldiers to use it. Like Jomini's "certain influence on the details of tactics", Dixon agreed the rifle would have an appreciable effect on battle, but that the effectiveness of the rifle would be forever crippled by the limited capacities of the average soldier using the rifle. Like the Napiers and Sir Howard Douglas before him, Dixon emphasized "the importance of always regarding *the rifle* and the *soldier* as inseparable in discussing the relative value of different arms, and by so doing we shall avoid forming, possibly, exaggerated notions of what may be anticipated from the infantry soldier in future warfare."[180] Whatever the theoretical capabilities of the rifle, the "class of men into whose hands the arm is placed" (i.e. the uneducated, largely illiterate urban poor from which the mid-19th century British Army drew its recruits) were inherently incapable of using a rifle effectively. In all fairness to Dixon, however, he did have something of an open mind. He confessed plainly, "I may have underrated the mental capacities of the average line soldier, but if so, I shall most willingly confess

---

[179] Trevalyan, 327
[180] Dixon, Matthew, "The Rifle--Its Probable Influence on Modern Warfare." *United Service Institution* Volume I (1858): 95

my error if it can be satisfactorily brought home to me."

Dixon also dismissed the remarkably forward-looking Belgian officer, Captain Francois Gillion, for predicting that modern weapons (such as the rifle-musket) would force infantry to fight more like skirmishers in open formations, instead of shoulder to shoulder in dense Napoleonic battalions. Dixon, and his conservative associates, spoke with some authority, as the military tactics and strategy of the preceding 200 years emphasized the maneuver of large masses of troops and the decisive culmination of battle being the concentrated *charge en masse*. What the advocates of the rifle were proposing was, in fact, a completely new method of combat based on theoretical capabilities that, at this time, had not been tested on a major European battlefield. The one major European battlefield they *did* have to study was Solferino, and it seemed to teach a contrary lesson. Lt. Colonel Dixon was prudently questioning the adoption of an untested novelty as his country's primary means of defense. "I cannot bring myself to believe or comprehend how it is possible to consider every line infantry soldier a marksman," Dixon said, "and one who is to be allowed to be constantly altering and adjusting his sight." He concurred with Napoleon III and Jomini, who would have agreed wholeheartedly with Dixon that "battles have never yet been won through the independent action of the individual soldier. The action of the mass operating as a mass at the decisive and critical moment will still, I think, be required to produce the greatest results."[181]

Perhaps the most fervent adherent of the French school to be published in Britain was the ardent Francophile, Lieutenant Andrew Steinmetz of the Queen's Own Light Infantry Militia. By the 1850s, the British Militia had evolved into a reserve force that, in a national emergency, would be tasked primarily with home defense. They would gather occasionally for drill, a few weeks each year. The ranks of the Militia were composed almost universally of lower class day laborers without steady employment, who appreciated a few weeks of guaranteed income

---

[181] Ibid., 116

during the periodic musters for drill. During the French war scares of the 1850s, the patriotic middle-class British men who felt compelled to do their part for the defense of England invented their own sort of reserve component – the Rifle Volunteer movement – rather than enlist in the Militia alongside the lower social classes. Militia officers were similarly seen as distinctly lower in social standing than officers in the regular Army.

Steinmetz was scathingly critical of the British system of musketry and neglect of the bayonet, but it should not be surprising that he was ignored and disregarded by his contemporaries. Indeed, he is only considered here because he provides a clear description of French shock tactics in the mid-1860s in English, and also because he was selectively quoted by Dr. Allen C. Guelzo, the Civil War historian of some prominence. In the book *Gettysburg: The Last Invasion*, Dr. Guelzo argues that the rifle-musket made little or no impact on the battlefield, and quotes this obscure militia lieutenant as an authority on musketry during the rifle-musket era. In fact, Steinmetz's military credentials are rather lacking. He was not a professional soldier, and was instead a barrister at law in Tower Hamlets and a prolific author of books across a spectrum of subjects. Born in 1819, Steinmetz became a Jesuit novice in the 1840s, subsequently left the Jesuits, and then wrote a controversial "tell-all" account of his experiences in 1848. He also wrote mildly controversial books on smoking, dueling, and his most successful work, a history of gambling. The ex-Jesuit lawyer was a practically elderly man of nearly 40 when, in the late 1850s, he was appointed an ensign in the Militia (most ensigns were usually in their teens). Steinmetz was sent to the School of Musketry at Hythe and did well, receiving a First Class certificate and being appointed as Instructor of Musketry in 1861 for the Tower Hamlets (Queen's Own Light Infantry) Militia. Shortly thereafter he departed for France and spent three months "studying the military organization of the French," including six weeks at the French military school at Vincennes. Impressed by what he observed, Steinmetz returned to England wholly infatuated with the French school and absolutely convinced of the superiority of the French system of war (with Solferino as its proof).

After his time in France, Steinmetz read two papers at the United Service Institution in 1862. The first was a summary of French cavalry tactics that Steinmetz, a Militia infantry officer assuming the mantle of the expert in all things military, vigorously urged the British cavalry to adopt. The second paper, the *Military Gymnastics of the French,* was submitted after the deadline, but the editors decided to print it anyways at the last minute. In *Military Gymnastics,* Steinmetz gushingly declares the molding of the French soldier "the utmost perfection possible at present." In no uncertain terms, Steinmetz wrote that "a system is now enforced in military tuition, which bids fair to make the French soldier and French officer the best in the world."[182] The first half of this paper describes the athletic gymnastics done to prepare French soldiers for taking the bayonet to potentially rifle-armed enemies. Such drills included quick-step marches to cross the dangerous space and rapidly scaling a parapet or taking a tower by *escalade* with ladders or even vaulting with poles. The French, he said, were "quite prepared" for the change of tactics that rifled arms and artillery might bring. "The emperor [Napoleon III] told them that such arms were dangerous only at a distance," Steinmetz wrote, "and that the bayonet would be, as it always was, the terrible arm of the French soldier. They will close in as quickly as possible, and the battle will be decided by the suddenness of the attack and the unabated vigour of practised hands, lungs, and legs."[183] When Steinmetz asked a French officer about the combined accuracy of his troops with the rifle-musket, the officer shrugged and explained that, "at Magenta, the battalion didn't fire a shot, but we made lots of prisoners."

Then Steinmetz proceeds to his scathing critique of the British

---

[182] Steinmetz, Andrew, " Military Gymnastics of the French." *United Services Institution Journal,* Volume 5 (1862), 370
[183] Ibid., 382

adoption of fire tactics in the second half. Lt. Colonel Dixon had argued that the "class of men" enlisted into the army were unable to use a "delicate" weapon like the rifle-musket; Steinmetz went further, insisting that the rifle-musket itself was inherently too difficult a weapon to use effectively at long range. He asked rhetorically if the British Army was not "somewhat bewildered in our estimate of the rifle?"

Although Steinmetz added "First Class Certificate, School of Musketry, Hythe," to his short list of military credentials, he dismissed his instruction at Hythe as a waste of time. "The regulations, the 'rules of fire,' of the target-ground and Hythe, the 'figures of merit' awarded and recorded so ostentatiously, are little better than vain delusions," Steinmetz boldly wrote. The sights on the rifle were "decidedly a great disadvantage" that "is always troublesome." Recall, if you will, that over eight years before Steinmetz wrote this, British soldiers *were* using their sights in battle, with perhaps the earliest example being at the Battle of the Alma, 20 September 1854. As briefly covered earlier, the historian of the Grenadier Guards recalls that a soldier of the battalion "quietly asked his captain to what distance he should set the sight of his Minié [rifle]."[184] In battle, Steinmetz argued, "the soldier cannot think of aiming at an enemy's feet, belly, or head, nor calculate distance. He directs his muzzle to the *breast* of an enemy -- in a word, right before him horizontally." Then he offers the line seized upon by Dr. Guelzo as evidence of the rifle-musket's general uselessness: "In fact, when surrounded by dust and smoke, how is it possible for the soldier to do more than fire horizontally?" In other words, use the rifle-musket in exactly the same fashion as the preceding smoothbore; level the piece and fire. This was the conclusion the French had reached, with fixed block sights. The rifle-musket, with its sights and great trajectory, was simply too "precise" a weapon for the common soldier to use effectively at long range. As Steinmetz concluded, "sure and

---

[184] Hamilton, Frederick William, *The Origin and History of the First or Grenadier Guards* (London: J. Murray, 1874), 192

terrible in practised and steady hands, it is certain that nineteen-twentieths of men will never be able to use it with perfect ease."[185]

The Militia lieutenant in his forties, with less than three years' experience as a junior officer in a unit whose part-time soldiers mustered for drill for a few weeks per year, even lamented the replacement of Brown Bess with the rifle. "It is an incontestable fact," Steinmetz unequivocally maintained, "that in certain respects the old smooth-bore musket was a better weapon than the present rifle, especially against the charge of cavalry."[186] He went on to advocate the removal of the precise rifle sights found on the Pattern 1853 Enfield, replacing them with the simple fixed block sight that the French used. Steinmetz did not stop with merely abolishing the sights, and proposed teaching a sort of snap shooting, from the hip, as more realistic for close quarters combat conditions than the long-range shooting ground at Hythe. Out of all the British sources on musketry from the mid-19th century, it is Lieutenant Steinmetz of the Tower Hamlets Militia that Dr. Guelzo chose to quote in his argument against the effectiveness of the rifle-musket. Ignoring the mountain of other primary sources, Dr. Guelzo cites this infatuated junior officer who rejected the British Army's institutional embrace of *firepower* as the primary infantry mode of combat, instead of the bayonet. Steinmetz readily supported the thesis that the bayonet was supreme and the rifle-musket made for a poor battlefield weapon; Dr. Guelzo elevated him as an authority on the subject. The present-day historiography on the rifle-musket appears increasingly to be a revisionist race to the bottom.

In spite of Steinmetz's crushing lack of military experience, his scathing essays offer an accurate summary of the French system. He was not wrong about the limitations of the rifle-musket, but like so many other critics of the rifle-musket who were not quite able to leave the bayonet or Brown Bess behind, Steinmetz ignored the performance of the rifle-musket (in trained,

---

[185] Steinmetz, 390
[186] Ibid., 385

experienced hands) in the Crimea and India. He remained firmly planted in the misconception that riflemen must always aim at individual human targets that they can see, never considering the possibility of rifles engaging "area" targets, a concept readily embraced by the School of Musketry (and reflected practically in the generous sizes of the targets used in the annual soldiers' qualification). He missed the subtle but groundbreaking transformation in battlefield musketry proposed by men like Colonel Wilford, and which would later come to be known as support by fire. Even so, Steinmetz was probably more realistic in his mental conception of modern battle than Colonel Wilford and his contemporaries. Steinmetz realized that the British Army's clean, regimented musketry instruction, with its tables and figures of merit, provided nothing of a so-called battlefield inoculation for soldiers. "Let us not for one instant believe that our target practice is to be the criterion of our achievements on the day of battle," Steinmetz wrote. "How will it be in the ranks at volley-firing or file-firing: the soldiers excited to the highest degree, cannon-balls decimating the ranks, shells and bullets whistling their infernal tune overhead (which no one forgets, having once heard it), surrounded by smoke, amid the groans of the dying and the shrieks of the wounded?"[187]

The battlefield was, indeed, a far cry from the sandy target-range at Hythe. Steinmetz advocated discipline and training in the French school, which in his estimation took a more realistic approach to the chaos and carnage of the battlefield. He also warned against using untrained volunteers: "undisciplined troops, brought under fire in masses, are always liable to ruinous panics, as has just happened in the American civil war."[188] Here Steinmetz sounds much like another soldier, albeit one with a lengthier set of credentials: the Prussian chief of the general staff, Helmuth von Moltke the Elder, who has been apocryphally quoted as describing the American Civil War as "two armed

---

[187] Ibid., 390
[188] Ibid., 394, referring probably to the First Battle of Bull Run

mobs chasing each other around the country, from which nothing could be learned."[189]

---

[189] Griffith, Paddy, *Battle Tactics of the Civil War* (Ramsbury, England: 1987), 21

10
THE RIFLE-MUSKET IN THE AMERICAN CIVIL WAR

So we come at last to the American Civil War, that great unrelenting period of brotherly carnage that has come to define the rifle-musket age in military history. The context of the American Civil War is painfully unique, especially for historians seeking to classify wars into neatly defined tactical or technological eras. Often cited as a candidate for the first modern war, or at least the first war of the industrial age, the mass-production in factories and use of telegraph, railroad, steamship and ironclad were truly a preface for the wars of the 20th century. One other aspect is equally true: it was certainly the *last* major war in which the overwhelming number of soldiers on both sides went into combat with little training (or no training whatsoever) in the use of the weapons they were issued. If the American Civil War was indeed the first war to use modern weapons, the lessons of that war must be unpackaged carefully and in the context of the volunteer civilian-turned-soldiers using them. In the middle 20th century, it was popular to compare the Civil War with the bloody trench-scarred fields of Flanders, or even call it a "dress rehearsal for World War I."[190] Many Americans have castigated European generals for ignoring the supposed lessons of the Civil War and the carnage caused by modern weapons, which had to be "re-learned" so to speak in the Boer War and especially in the First World War. This view has (thankfully) generally been replaced among serious historians with the modest proposal that the Civil War was among the last Napoleonic wars, and a rather incompetently fought one at that, when evaluated across a spectrum of various factors. In this vein, it is currently fashionable in the contemporary historiography to question the old narrative that the adoption of modern weapons like the rifle-

---

[190] Ibid., 20

musket made the Civil War so much bloodier, longer, and transformative than wars before it. There is a tendency to go even further and assert that the rifle-musket, by virtue of its failure to decisively change the face of battle in the Civil War, was not a transformative or revolutionary weapon (let alone the first modern infantry weapon).

Since Paddy Griffith inaugurated this much-needed reconsideration in his controversial *Battle Tactics of the Civil War*, historians have largely come to argue that the rifle-musket was not a modern weapon, nor did it usher in any significant change to Napoleonic tactics. Griffith was ignored and even maligned by Civil War historians for over a decade after the publication of *Battle Tactics*, but he has experienced a recent vindication. His primary argument -- that Civil War battles were fought at smoothbore ranges in spite of the rifle-musket -- has been finally accepted in the mainstream. Modern historians have recently gone even further, making arguments for average battlefield ranges of less than 100 yards. They have, alas, thrown the baby out with the bathwater. These historians are, I believe, far too broad in their ultimate conclusions about the rifle-musket. They have pronounced that the rifle-musket itself was a flawed weapon, and it could not have influenced warfare any more than the smoothbore musket, except in peripheral areas such as skirmishing. A more accurate assessment is that in the hands of an untrained American volunteer soldier in the Civil War, the rifle-musket was not a modern weapon that influenced battlefield tactics. However, in the hands of soldiers who received even minimal musketry instruction, the rifle-musket was, as we have seen, used at long range on dozens of battlefields and was devastatingly effective. Could it be that Steinmetz and von Moltke were, to a certain extent, correct in their evaluation of the American Civil War as mere "undisciplined masses" with scarcely any training whatsoever, resembling more an "armed mob" than a mid-19th century army?

Standing on top of a mountain of scholarship today, it can be fairly and unequivocally stated that the average American Civil

War soldier knew very, very little about the military application of musketry. His officers were scarcely any better. The entire United States Army, prior to secession, represented scarcely 18,000 officers and men. There was no foreign threat of imminent invasion by a large force of well-trained European troops; the US Army was primarily occupied with the pacification of Native Americans on the frontier. Volunteers surged into the armies of North and South after the crisis of secession, and by March 1862 they had 500,000 men in the field. Practically none of them knew anything about warfare. Even if there were zealous advocates of military musketry in high leadership positions, it was impossible to conduct musketry training due to the lack of rifle-muskets and expendable ammunition. There were only 36,000 rifle-muskets in the United States when the Civil War began.[191] Arsenals, Union and Confederate, could scarcely meet the demand for cartridges (in a dizzying number of variations and calibers) for immediate use in combat, let alone a surplus for musketry practice. To simply send the 500,000 soldiers of the March 1862 armies through the British Army's standard annual musketry qualification course would have required 55 *million* cartridges. Sustaining this level of musketry training was not an issue in Britain, where government and commercial manufacturers had the infrastructure to produce many millions of cartridges a year. In the war-torn American states, there was scarcely enough ammunition for fighting and certainly none to "waste" at such whimsies as firing at targets 900 yards away!

The combination of these factors -- inexperienced soldiers without training, the lack of rifle-muskets to arm them with, and the lack of ammunition to expend in training -- led to the inevitable result: American Civil War soldiers were unable to utilize the rifle-musket's capabilities. Instead, they used it like a smoothbore musket and fired at point-blank ranges that did not require any knowledge of trajectory or use of elevation on the sights. Paddy Griffith correctly noted that "there is little in the literature to suggest that the average Civil War infantry regiment

---

[191] Ibid., 32

even began to judge distances or set sights accurately for battle."[192] On those remarkably rare occasions that a Civil War regiment did conduct target practice, it was actually something to write home about, as soldiers' letters home and diaries attest. Griffith describes a "serious lack of target practice in the armies of both sides," and when a unit did conduct some rudimentary form of musketry training, the soldiers "regarded it as a highly exceptional event."[193]

Many American officers reached the same conclusion about the rifle-musket in the hands of Civil War soldiers, who in many ways resembled the highly enthusiastic yet woefully untrained volunteers of the *Risorgimento*. Brigadier General Stephen Vincent Benet, a U.S. Ordnance officer during the Civil War and later the 8th Chief of Ordnance, was a staunch advocate of musketry instruction after the war. In a letter from August 1882 to General William T. Sherman, who was then the general commanding the U.S. Army, Benet recalled the unenthusiastic approach to military marksmanship before and during the Civil War. In the tiny antebellum army, "the capabilities of the new rifle, as to range and accuracy, was entirely beyond comprehension." Officers spent years (even decades) as lieutenants, waiting for promotions in the small army; this meant that every antebellum officer in the grade of captain and higher had been "brought up under the use of the smooth bore musket" and were "slow to abandon the ideas imbibed by the old smooth bore." The inevitable consequence was that up to the outbreak of the Civil War in 1861, "little or nothing had been done in the Army toward systematic instruction in rifle practice." When the war broke out and the armies swelled with millions of civilian volunteers, there was no musketry instruction waiting for them and the only people with any experience were the Regular Army officers and men, still "imbibed" with the ideas of the old smoothbore. "Thus," Benet unsurprisingly concludes, "the value of the rifle as to accuracy

---

[192] Ibid., 88
[193] Ibid., 87

was in a great degree lost for want of proper training to the soldier." Even more appallingly, "tens of thousands of men fired in battle their first shots with the Army rifle."[194]

It should be no surprise, then, that the American Civil War failed to produce the battle-changing revolution in tactics that the first supposedly modern, industrial war was supposed to do. Griffith readily states that there was "an almost total lack of target practice" and a "meager supply of cartridges" available, but concludes in the same paragraph that "even with these wonderful new weapons [rifle-muskets], in fact, it remains doubtful that a genuine revolution in firepower had actually occurred."[195] Griffith repeats this cornerstone of his thesis in his conclusion, definitively stating that "the idea that the rifle musket revolutionized tactics [is] demonstrably false."[196] It should astonish nobody that soldiers sent into battle with an "almost total" absence of weapons training failed to use their weapons effectively. Put very simply, the American Civil War soldier did not know how to use the rifle-musket, because his officers did not know how to use the rifle-musket, and his generals fought the war in the fashion of the old Napoleonic smoothbore, for lack of a reasonable alternative (and not out of tactical or strategic ignorance). Dr. Hess is correct in saying "weapons are tools, and their level of effectiveness depends on how they are used."[197] Civil War soldiers, mostly in their teens or early twenties, left their farms and shops and factories to volunteer: they were haphazardly uniformed and equipped, taught drill largely by officers who were scarcely proficient themselves, and sent into battle with a weapon that in many cases they had never even fired before. Even as the war dragged on for years, musketry instruction was exceptionally rare in Civil War armies. Imagine

---

[194] Benet, Stephen, "Rifle Target Practice in the Army." *Army and Navy Journal*, Volume 20 (1883), 176
[195] Griffith, 90
[196] Ibid., 189
[197] Hess, Earl, *The Rifle Musket in Civil War Combat* (University Press of Kansas: 2008), 119

taking a 18-year-old barista at Starbucks, putting a uniform and body armor on him, handing him an M4 carbine, and sending him almost directly into combat in a company of similarly untrained soldiers; when they flatly fail to "revolutionize warfare" on the battlefield, can we honestly attribute this result to the complexity and inherent difficulty of correctly using the M4 carbine? This is exactly what Civil War historians are saying, when they equate the contextual failure of American Civil War soldiers to use the rifle-musket effectively to the apparent universal failure of the rifle-musket to revolutionize battle. The *same* weapon misused by Civil War soldiers -- and ironically, in more than a few cases, the *exact* same weapon -- was used effectively by British, French, and East India Company troops in the Crimea, in India, and elsewhere, who were taught how to use their arms.[198] As Dr. Hess observes, the effectiveness of the weapon depends on how it is used.

The deck seems to have been stacked against the rifle-musket in the American military context from the start. Ironically, the pre-war U.S. Army was reluctant to even use the rifle-musket solely as a rifle, and looked into ways to continue using it in the fashion of the smoothbore. Following extensive experiments conducted by officers the U.S. Ordnance Department from 1853 to 1855 at Harpers Ferry and Springfield Armory, the Model 1855 rifle was adopted as the primary infantry weapon for the U.S. Army. This was a fine rifle with a progressive depth rifling of essentially the same kind as the Pattern 1853 Enfield. The Model 1855 rifle of .58-inch caliber was sighted out to long range with a complex elevating rear sight. Its capabilities were squandered in the antebellum army. One particularly revealing example is the "fire of three balls" somewhat gleefully recommended by the Ordnance officers. It was seriously proposed that the brand-new

---

[198] More than three quarters of all Pattern 1851 Minié rifles made were declared surplus in the early 1860s and sold to arms dealers for American import, where they were disliked by Civil War troops and often called "rifled Brown Besses". Today, these scarce rifles are more commonly encountered in the United States than in the UK, for this reason.

Model 1855 rifle could effectively fire *three* round balls of .525-inch diameter out to 200 yards. Essentially, the proposal was to use the new rifle precisely like the old smoothbore, but owing to the greater strength provided by the steel rifle barrel, it could safely discharge three round balls without risk of bursting. Firing three balls from the older iron smoothbore musket barrel was extremely unsafe. The undersized balls shot from the Model 1855 rifle had no means of engaging the rifling, and so they were blasted out with no more accuracy than the old smoothbore. A year after the rifle-musket had cleft the Russian armies in Crimea like the "destroying angel," the U.S. Army followed the European example and adopted a brand-new rifle-musket. Revealingly, the *first* serious suggestion of tactical employment of the new rifle was to use it like a glorified smoothbore, for firing round balls at close range. I have not found any evidence that the three-ball ammunition was ever actually used, probably because once a three-ball load was fired, the rifle barrel would have to be carefully cleaned before a regulation rifle bullet could be loaded.[199]

A handful of U.S. Army officers did attempt to sound the alarm, as they keenly observed the widening gulf between capabilities of the U.S. Army and the British and continental armies. With the adoption of the Model 1855 rifle, an Army circular dated March 15, 1856 "respecting practice with small arms" was issued, directing commanders to conduct firing practice. This "practice" was haphazard at best, in the absence of any regulation. To fill the void, the Army formed a board of officers in August 1857 to "draw up a 'System of Target Practice with Small Arms'" and directed Captain Henry Heth, an officer on the board, to write the new regulation. Captain Heth spent the next few months studying French and British books on musketry instruction. He eventually came to the unsurprising conclusion that a complete system like the British and French had would be impossible to replicate in the U.S. Army, because of the lack of

---

[199] *Reports of Experiments with Small Arms for the Military Service by Officers of the Ordnance Department* (A.O.P. Nicholson, Washington DC: 1856), 111

an established school like Vincennes and Hythe for training up a cadre of skilled instructors. Heth's final product, *A System of Target Practice: For the Use of Troops when Armed with the Musket, Rifle-Musket, Rifle, or Carbine*, was adopted as the US Army regulation by order of Secretary of War John B. Floyd in March, 1858. There was no pretense that Captain Heth's work was at all original; the title page plainly confessed the work was "prepared principally from the French by Henry Heth." In his introduction, Captain Heth described his manual as "chiefly a translation from the French." Lamenting the lack of an American "school of musketry," Heth said "the French system would have been recommended, with but little or no change, had we in our service schools of instruction similar to theirs." Ultimately, Captain Heth could only admit that he "does not claim the credit of presenting to the army any thing new, but only a digest of what has already been practiced, with great success, by both the English and French."[200] The new regulation was not received well by the Army's senior commanders, "imbibed" with the ways of the old smoothbore. A few junior officers passionately advocated for musketry instruction, but Captain Heth's new regulation was largely ignored.

One of the few advocates for musketry training was Lieutenant Cadmus Wilcox, whose 1859 *Rifles and Rifle Practice* is the only serious American work on the subject that is comparable with the myriad of contemporary British works.[201] Lieutenant Wilcox, of the 7th Infantry Regiment, was exasperated by the lack of musketry instruction in the United States. As an expert on rifles and rifle shooting, Wilcox realized that the rifle was a new kind of weapon that was destined to radically change the

---

[200] Heth, Henry, *A System of Target Practice: For the Use of Troops when Armed with the Musket, Rifle-Musket, Rifle, or Carbine* (Henry Carey Baird, Philadelphia: 1858)

[201] Like his contemporaries, Lieutenant Wilcox freely confessed that his work was not entirely original, borrowing heavily from early French works on musketry, especially the mathematical explanations of trajectory.

battlefield. He was in perfect agreement with the officers of the British Army's School of Musketry at Hythe, that soldiers would require specific and focused training in order to effectively use the new weapons. "A rifle," Lieutenant Wilcox wrote, "whatever may be its range and inaccuracy, in the hands of a soldier unskilled in its use, loses much [of] its value, hence the necessity of giving the most thorough practical instruction."[202] In this, Lieutenant Wilcox would have wholeheartedly agreed with Dr. Hess: the effectiveness of a weapon depends on how it is used. Wilcox went on to advocate for "the necessity of creating schools specifically for the purpose of teaching the soldier the art of firing." He urgently reminded the reader that "such schools are now general in Europe." Predicting that the rifle-musket would be a four-fold improvement over the smoothbore musket (a statement General Havelock certainly would have agreed with), Wilcox emphasized "the necessity, in order to secure the full effect of the arm, to have thorough system of instruction in target practice; every infantry soldier should be so instructed before he enters his battalion."[203]

Although Heth's *A System of Target Practice* remained the official U.S. Army regulation for musketry instruction from 1858 through the Civil War, there is no evidence whatsoever that any serious attempts were made to abide by it. Instead, the massive armies of volunteers were immersed in drill, usually doing their best to follow brevet Lt. Colonel William Hardee's 1855 *Rifle and Light Infantry Tactics*. It was based on the much older French tactics of the *chasseurs à pied*, which had ironically by 1855 fallen out of favor even in France. Hardee incorporated the *pas gymnastique*, increasing the rate of march from Scott's stately 90 to 110 steps per minute to the 165 to 180 steps per minute of the chasseurs. Hardee assumed the soldiers in these formations would also be well trained in fencing with the bayonet, and also skilled rifle shots, like the French chasseurs. At any rate, Hardee's

---

[202] Wilcox, Cadmus, *Rifles and Rifle Practice: An elementary treatise upon the theory of rifle firing* (Van Nostrand, New York: 1859), 238
[203] Ibid., 243

Tactics only covered unit maneuvers up to the battalion level. For larger formation maneuvers, the U.S. Army used a translation of the very early French infantry *Ordonnance* manual that was revised 1831 by Brenier and Curial, and immediately "faithfully translated" by the U.S. Army's General Winfield Scott in his 1835 *Infantry Tactics*. A year after the battles of Alma, Balaclava, and Inkerman, where trained British soldiers with rifle-muskets had demonstrated the capabilities of the new weapon against Russian infantry, cavalry, and artillery in detail, the U.S. Army doubled down on essentially the same Napoleonic shock tactics unsuccessfully employed by the Russians in the Crimea. The object of Civil War tactics remained strictly Napoleonic, with the goal of turning the enemy's flank; American Civil War soldiers were drilled in these formations, to the certain horror of men like Colonel Wilford, like robotic unthinking automatons. Ammunition was expensive and scarce, but drill was cheap and Civil War volunteer soldiers consumed reams of paper writing home to complain about the endless hours, day after day, or unrelenting drill. I agree with Paddy Griffith that the American Civil War soldier became, after several years of service, the approximate equal of a European soldier in proficiency in drill. In the smoothbore era, battles were won by maneuvering and turning flanks, putting units in position to deliver fire "with effect" through massed musketry, where the proficiency of the individual or accuracy of his fire was the last of anyone's concern. Accordingly, the concept that the individual soldier should be trained to effectively use the rifle-musket combat multiplier was almost completely foreign. American Civil War armies trained for, and fought, the battles of a previous era of war.

Any effort to conduct musketry instruction among Civil War armies appears to be exceptional, especially prior to 1864. A British officer in Canada, Lieutenant Sydney Herbert Davies, resigned his commission and traveled south to volunteer for the Confederate Army. He requested a major's commission, since after all he was a professional British officer, "in possession of a first class certificate as an instructor of musketry and am not ignorant of warfare." The Confederates apparently had to little interest in musketry instruction that they could unenthusiastically offered this Hythe-trained musketry expert a commission as a lieutenant.[204] One such exception was General Patrick Cleburne's division in the otherwise stolidly old fashioned Confederate Army of Tennessee. The Irish-born Cleburne had been a corporal in the British Army before emigrating to the United States and enlisting as a private in the Arkansas militia as secession loomed. As Cleburne rapidly rose through the Confederate ranks, he brought a distinct outsider's perspective of the Army (as opposed to the aging West Point-trained Regular Army officers "imbibed" in the ways of the smoothbore musket). Cleburne was one of very few officers on either side that aggressively sought to train the *individual* soldier as well as drill the collective formations. Facing the fully mobilized industrial might of the North, with superior numbers of Union troops, Cleburne was looking for a combat multiplier. By 1863, he had instituted target practice (albeit with a very small allocation of ammunition) and dispatched an officer on his staff, Major Calhoun Benham, to publish a manual of musketry instruction. Major Benham traveled to Richmond and ultimately copied, nearly word for word, the British Army's 1859 *Regulations for Conducting Musketry Instruction of the Army*.[205]

Benham's little book was adopted as the Army of Tennessee's manual for musketry instruction, but this worthy effort was crippled from the start due to the lack of cartridges for practice shooting. The British Army's "Preliminary Drill" containing such practices as Judging Distance, Position Drill, and Theoretical Principles was copied nearly verbatim in the Confederate manual. The actual "Target Practice" chapter, which encompassed the British soldier's annual firing of 90 to 110 rounds, was deliberately left out. "It is not thought worth while to add the

---

[204] Foreman, Amanda, *A World on Fire: Britain's Crucial Role in the American Civil War* (New York: Random House, 2011), ii
[205] Powell, David, *The Chickamauga Campaign* (Savas Beatie: 2014), 594

directions for practice," Benham explained. "The situation of our armies, and the economy necessary in ammunition, render it impossible to practice at the target to any great extent."[206] In the introduction to Benham's *A System for Conducting Musketry Instruction*, he confesses "this book is in some parts a copy, in others a mere analysis" of the British Army's *Regulations,* and that the entire firing practice had been omitted "for the reason that it would be inconvenient, if not impractical, to teach them to an army in the field."

European observers were not impressed by the manner with which the Civil War armies utilized the rifle-musket. Even Lieutenant Steinmetz, with his dim view of the importance of individual marksmanship training, was appalled by the misuse of the rifle-musket in American hands. Lieutenant Walker, of the School of Musketry, declared that it was "notorious that the mass of the armies, especially that of the North, is merely a crowd of undisciplined, untrained volunteers -- but yesterday plying the quill, the plough, or the shovel." Walked blamed the lack of distance estimation training for the inability of the Civil War armies to make "to some account the beautiful rifles" they were armed with. "What would not the contending armies in America give for a thorough knowledge of judging distance?" Walker asked. "Nothing can afford a stronger proof of the necessity of judging distance than the comparatively slight effect of the introduction of the rifle into the armies in question."[207] Another British officer, George Francis Robert Henderson, wrote in his work *The Campaign of Fredericksburg* that Union "infantry soldiers, as a rule, marched fairly well, were brave and stubborn in the field, and patient under reverses; armed with a serviceable and long-ranging rifled muzzle-loader, they were but indifferent marksmen, and often extremely careless of the condition of their

---

[206] Benham, Calhoun, *A System for Conducting Musketry Instruction* (Richmond: 1863), 28
[207] Walker, 126

arms."[208]

In the only full-length published work on the subject of the rifle-musket in the American Civil War, Dr. Hess's *The Rifle Musket in Civil War Combat* is most valuable for the primary source documentation of Civil War soldiers themselves describing the ranges and ammunition expenditure of infantry firefights. In addition to this excellent index, the chapters on skirmishing and sniping are the first (to my knowledge) specific considerations in any published work. I believe Dr. Hess is, however, *too* critical of the rifle-musket, and too eager to exaggerate its inherent shortcomings. His thesis seems to be summarized by a succinct statement in his introduction: "It [the rifle-musket] did not revolutionize warfare."[209] As the most popular and widely-cited work on the subject, Dr. Hess's *The Rifle Musket in Civil War Combat* deserves a thorough and careful treatment here.

On the first page of his book, in the introduction, Dr. Hess described the rifle-musket as "largely untested in the hands of ordinary soldiers on a real battlefield," and a "new weapon as yet unburdened by the realities of actual use in war."[210] These statements would be true only if prefaced by *American* soldiers, and *American* war; by 1861, the rifle-musket had been used effectively (and ineffectively) for years on battlefields around the world. It becomes quickly apparent that Dr. Hess makes universal assertions about the rifle-musket itself, but only in the context of the American Civil War. It would be unfair to expect Dr. Hess to spend much time on the Crimean War and other antebellum rifle-musket conflicts, but while he correctly acknowledges the rifle-musket as being "widely successful" during the Crimean War, he asserts that the British contingent was only one-third armed with the new "Enfield" rifles. The British Army in the Crimea was actually almost completely armed with the Pattern 1851 rifle; only one out of four British divisions

---

[208] Henderson, George Francis Robert, *The Campaign of Fredericksburg Nov. -- Dec., 1862* (Kegan Paul Trench & Co, London: 1886), 19
[209] Hess, 3
[210] Ibid., 1

that landed in Crimea was armed with smoothbores, and these were quickly replaced with rifles in due time. Dr. Hess's only mention of the use of the rifle in the Crimean War is for sharpshooting during the Siege of Sevastopol. He found nothing to learn from the Indian Rebellion of 1857 either, citing the "limited" use of rifles.

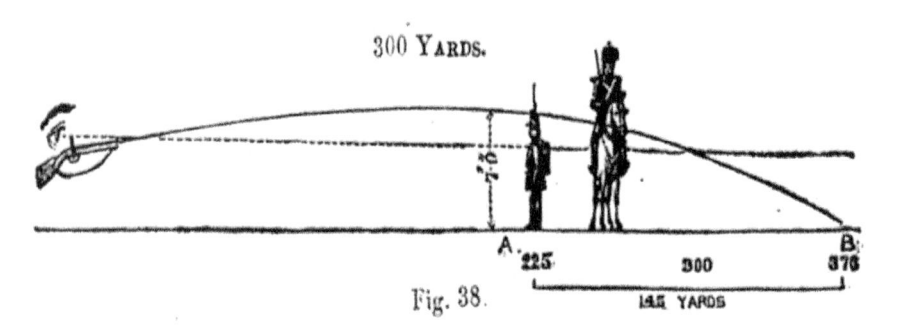

Fig. 38.

Dr. Hess's entire thesis is undermined by his profoundly incorrect understanding of the trajectory of the rifle-musket. He describes the 300-yard trajectory of a rifle-musket as having a "killing zone" (i.e. dangerous space) of "only about 75 yards long," with the equally incorrect conclusion that for "nearly half the 300-yard range, enemy troops would be untouched by the balls."[211] The actual dangerous space is over *twice* what Dr. Hess claims, at 145 yards for the P1853 Enfield rifle (with essentially an identical span for the M1861 Springfield and other contemporary rifle-muskets). The only place that troops would be "untouched by the balls" in this trajectory is an uncomfortably narrow span in the middle of the trajectory where the bullet reaches its zenith. Dr. Hess points to this precarious zone of invulnerability to highlight the difficulty of accurately shooting a rifle-musket, but does not explain why soldiers would set their sights at 300 yards to fire on enemies at 150 yards.[212] Cavalry,

---

[211] Ibid., 2
[212] It is possible that Dr. Hess refers to certain models of European rifle-muskets, many of which had fixed, nonadjustable "block" sights set to

incidentally, would remain in the dangerous space for the entire flight of the bullet, from the muzzle until the bullet's first graze of the ground at 370 yards, as the period engraving indicates.

Unfortunately, Dr. Hess seems to build his entire argument upon an incorrect understanding of the rifle-musket's trajectory and a dangerous space less than half of what it actually is. He also tends to overlook the fact that, with sights set at 300 yards and using proper shooting fundamentals, the shooter would *hit* his intended target 300 yards away (a feat impossible with any smoothbore). Because the bullet rises to a height of 7 feet on a 300-yard trajectory, Dr. Hess considers this a serious defect in the weapon that created what he calls a "safe passage" for enemy soldiers in the place that the bullet was higher than their heads. "Bullets would fly over the heads of many opponents," Dr. Hess wrote, but offers no explanation of why a soldier would be firing at enemies 300 yards away, having set his sights for 300 yards, while paying no attention to the enemy soldiers standing 150 yards away. Based on this incorrect trajectory, the presumed halved size of the dangerous space, and the enormous boon to the enemy created by the "safe passage" beneath the zenith of the bullet's flight, Dr. Hess builds his argument on the use of the rifle-musket in the Civil War.

Ostensibly because of the uselessness of firing at longer ranges that create convenient "safe passages" for enemies to approach unscathed by bullets, Dr. Hess's conclusion was that Civil War soldiers rarely (if ever) adjusted their sights, and fired at an average range of 94 yards.[213] This is substantially lower than

---

approximately 300 yards. Untrained, green troops who did not know any better could indeed send their bullets over the heads of enemies 150 yards or closer, if they did not know that they needed to aim low with the fixed 300-yard sights. Several German sources, both Austrian and Prussian, complain about this. Most rifle-muskets used in the American Civil War, however, had 100-yard sights (such as the M1861 Springfield and P1853 Enfield), and Paddy Griffith emphasizes that the evidence does not support the theory that American Civil War soldiers adjusted their sights beyond the 100-yard setting very often, if hardly at all.

[213] Ibid., 108

Paddy Griffith's average of 141 yards. These range estimations assume that the soldiers who later wrote about their battle experiences were accurate in their judgement of the distance between them and the enemy. It is safe to conclude, however, that Civil War musketry in battle was conducted at very close range. The question is why? All of the evidence points to the American Civil War being fought clumsily as the last Napoleonic war (although the Civil War is so unique that it even calling it "Napoleonic" is, to some degree, generous), with Civil War armies simultaneously unable and unwilling to harness the capabilities of the new rifle-musket. Tactics and technology in Europe had progressed far beyond the levels demonstrated by Civil War armies, leaders, and soldiers.

As Andrew Haughton argues in his excellent *Training, Tactics, and Leadership in the Confederate Army of Tennessee*, Civil War armies like the Army of Tennessee in his study "were, in effect, still using tactics developed for smoothbores."[214] Even when equipped with the rifle-musket they used the new weapon *like* a smoothbore, due to the logistical and practical impossibility of providing weapons training for enormous numbers of volunteers in the middle of a full-scale war. For most of the war, "the objective had always been to turn the enemy's flank," Haughton argues, "rather than progress beyond the opponent at the tactical level."[215] In other words, there was never any real attempt to make the individual soldier a "force multiplier" by training him to exploit the new theoretical capabilities of the weapon in his hands. There was tactical innovation in the Civil War, as Paddy Griffith and others have demonstrated, but these innovations themselves emerged only in a style of warfare that was already obsolete. The mark of modern war, and the modern soldier, is that great emphasis is now placed on the training, quality, and proficiency of the individual soldier, instead of the proficiency of massed formations to execute battlefield movements. By 1861,

---

[214] Haughton, Andrew, *Tactics, Training, and Leadership in the Confederate Army of Tennessee* (Routledge, New York: 2000), 177
[215] Ibid., 119

the armies of Western Europe had already turned this corner; British officers were discussing embryonic fire and movement tactics, and individual soldier proficiency with his assigned rifle had been successfully implemented for nearly a decade.

Only after years of reverses, in the apocalyptic summer of 1863, did the Confederates (at Cleburne's urging) belatedly initiate any attempt to make the rifle-musket a force multiplier by making the Confederate soldier *better* at the individual level than his Union counterpart. Major Benham's manual was printed in small numbers in September 1863, and parsimonious allocations of cartridges were provided for sporadic target practice. It was far too little, and way too late, to really matter.

11
THE AUSTRO-PRUSSIAN WAR 1866

Less than fourteen months after General Robert E. Lee surrendered his Army of Northern Virginia at the end of the American Civil War, and exactly three years to the day after the glorious Jominian failure of Pickett's Charge at the Battle of Gettysburg, the Prussians fought the Austrian Empire at the Battle of Königgrätz. The battle fought in the modern-day Czech Republic on July 3, 1866 is possibly *the* clearest, most decisive illustration of a paradigm shift in modern military history. As such, the Austro-Prussian War is an incredibly rich (and frequently overlooked) subject to study, especially for historians interested in paradigm shifts and, for our purposes, the first modern infantry weapons. The proximity of the Austro-Prussian War to the American Civil War invites comparisons, and military writers had begun to contrast the seven-week war in Europe with the four years of civil carnage in America even before the smoke had cleared from the fields around Sadowa.

Ironically, the side armed with the rifle-musket (the Austrian Empire) was utterly crushed at Königgrätz. The Prussian Army had adopted the breechloading Dreyse *Zündnadelgewehr* (needle-rifle), capable of a sustained rate of fire at least twice that of the most efficient soldier with a rifle-musket, and it could be loaded and fired from the prone. Prussian victory at Königgrätz is usually attributed to the needle-rifle, or at least to a combination of brilliant Prussian operational concentration of forces enhanced by the needle-rifle. Contemporary popular history considers the Prussian victory an inevitable, obvious result of a battle between one army with "modern" weapons and the other with obsolete, inferior arms. Many historians, including Dr. Hess, point to Königgrätz as evidence of the rifle-musket being the last of the pre-modern weapons, and present the breechloading rifle as the first modern infantry weapon. Therefore, in this book that argues that the rifle-musket was the first modern infantry weapon, there is no alternative but to deal with the events and armaments of 1866 comprehensively.

Instead of proving that the rifle-musket did not revolutionize warfare and was not the first modern infantry weapon, Königgrätz demonstrates the first application on a European battlefield of the principles and tactics developed *for* the rifle-musket and the modern soldier. The Prussian infantry used a breechloading rifle, but they were not the first to conceptualize defeating an enemy with pure firepower delivered accurately, rapidly, and overwhelmingly at distances beyond the range of the old smoothbore. Some of these concepts were being aggressively advocated for by the cadre of instructors at the British Army's School of Musketry almost a decade before the decisive clash in 1866. In fact, it was the Prussians who were the latecomers to *Feuertaktik*, the wonderful *Zündnadelgewehr* notwithstanding. As late as 1864, in the Second Schleswig War, Prussian tactical doctrine retained shock and the bayonet charge for clinching victory, instead of firepower. By 1866, the School of Musketry had been operating for 14 years, and hundreds of thousands of British soldiers had been shooting their annual qualifications, expending 90 rounds firing at targets from 150 to 900 yards, year after year. Seven years earlier, the British Army had unambiguously pronounced in the regulations that a soldier who cannot shoot is not merely useless, but an actual encumbrance to his unit. Institutional reformers had dissolved the distinctions between light and line infantry, making every British infantryman a rifleman, and doctrinally establishing in regulation that the British Army would try, at least, to defeat its enemies by pouring a hail of lead onto them. They had even come to accept the reality that the vast majority of bullets would fail to hit an enemy soldier, and correspondingly accepted the financial and logistical challenges of a greater expenditure of ammunition in both training and in battle. The Royal Laboratory at Woolwich, working in tandem with doctrinal innovators like General Hay and the School of Musketry, developed the "perfected" rifle-musket cartridge for the P1853 Enfield that could be loaded with great ease compared to contemporary cartridges. This new ammunition, introduced in 1859, allowed the P1853 Enfield to be fired virtually indefinitely, and opened the door to new developments like the embryonic "fire and movement" tactics that were only possible with trained troops equipped with a weapon capable of sustained long-range fire. All of these developments had been made with the rifle-musket as the universal infantry weapon of the British Army. They did not require a breechloading rifle to commence the paradigm shift. When the British Army first adopted a breechloader in 1866 (the Snider-Enfield, with modern centerfire ammunition), the new rifle merely inherited a mature process, already well underway, that the rifle-musket had initiated.

Prussia did not, however, look to the British Army's emphasis of trained riflemen as a model to emulate. Instead, the Prussians independently reached similar conclusions on their own, although a few years after the British had incorporated fire tactics into their doctrine. With the added variables of the rapid-firing Dreyse needle-rifle and the highly efficient Prussian General Staff, the Prussians took *Feuertaktik* well beyond where the British left off. While the 19th century historiographical "Great Man" model is fallen entirely out of academic favor, there was arguably one Prussian "great man" whose influence forcefully wrought the revolutionary changes in Prussian strategy and tactics: Helmuth von Moltke. His legacy remains strong today, and perhaps stronger than it even was during the Great Wars, now that the U.S. military has increasingly adopted Moltke's "mission command" doctrine since the 1980s. As Dr. Geoffrey Wawro observed in his excellent book *The Austro-Prussian War,* Moltke was the only Prussian in a position of influence in the pivotal 1860s to defy the standing column-and-shock tactics of the day, and recast Prussian infantry tactics around firepower.[216]

As chief of the Prussian General Staff, Moltke studied and absorbed the lessons of 1859, where the victorious Italo-French armies defeated the Austrians at Solferino and Magenta using enhanced variations of Napoleonic column and shock attacks. At first glance, the battles appeared to be a decisive vindication of Jomini and the French school of the bayonet, although a closer examination revealed that the vast majority of staggering number of casualties had been caused by bullets, not bayonets. Neither side was fully armed with rifles, and with the exception of a few highly-trained specialized units, practically none of the soldiers engaged at Solferino or Magenta had received anything close to the British Army's level of marksmanship training. Europe doubled down on Jominian shock tactics; only the British and Prussians dissented. Colonel Wilford, in a lecture given in November 1859, emphatically insisted that "the capabilities of long-range musketry were not fully tested on the Italian plains" and urged his audience to avoid being swept up in the fervor around the French shock tactics.[217] Moltke similarly grew to reject the Jominian "laws of war" and increasingly began to understand that the telegraph and railroads offered new strategic options that Napoleon could not have dreamed of. The improved industrial and logistical capabilities of nation-states also meant that armies were much larger than Napoleon's, and the Jominian principles that might have worked for an army of 100,000 would not answer for armies of 300,000.

Prussia's move towards *Feuertaktik* was accelerated after the brief Schleswig War of 1864 against Denmark. In 1865, Moltke wrote an illuminating essay, *Remarks of the Year 1865 on the Influence of Improved Firearms in Tactics,* distilling his observations from the Schleswig War. Before 1864, the Prussian Army had waffled between shock tactics and fire tactics. Field Marshal von Wrangel, who initially commanded the Prussian forces against Denmark,

---

[216] Wawro, Geoffrey, *The Austro-Prussian War* (New York: Cambridge University Press, 1996), 12
[217] Wilford, 59

called Moltke's fire tactics "uncontrollable" and "dishonorable." A French officer observing Prussian maneuvers was disgusted by the fire tactics employed, declaring that "Prussia is compromising the military profession."[218] But Moltke was convinced, and in his 1865 *Remarks* observed that "the short campaign against Denmark has proven the excellence of our weapons," but he warned that the poorly equipped and vastly outnumbered Danes "did not test our weapons in battle against an equally well-equipped enemy." He refused to follow Jomini's example and lay down any rules or principles that he was highly skeptical of. "It is broadly perceived that the substantial improvement of firearms will result in a change in the style of fighting [*Fechtart*] of troops," Moltke wrote, "but there is insufficient combat experience [*Kriegserfahrung*] to establish any rules."[219]

Much praise and credit for victory in the Austro-Prussian War has been heaped upon the Prussian needle-rifle. In fact, it is often cited as *the* principle reason Austria lost the war. This is a gross oversimplification and by 1866 the Dreyse needle-rifle was certainly not the unquestionably superior weapon that popular historiography has made it out to be. While the *Zündnadelgewehr* contributed substantially to Prussian victory in 1866, they probably could have won without it. Excellent staff work made a rapid mobilization possible; an efficient railroad system quickly transported Prussian reserves to the theater of war. Operationally, Moltke achieved an envelopment of the Austrians with three Prussian army groups at Königgrätz, and Prussian commanders at all levels often seized the tactical initiative to exploit opportunities. Trained Prussian soldiers delivered accurate rifle fire, and demonstrated the benefit of extensive marksmanship training. Meanwhile the Austrians mobilized slowly, moved cumbrously, and rejected *Feuertaktik* completely, committing to French-style shock column assault tactics. If the Prussian Army had been armed with a quality Minié rifle instead

---

[218] Wawro, *The Austro-Prussian War,* 23
[219] Moltke, Helmuth Graf von, *Moltkes Taktisch-Strategische Aufsätze auf den Jahren 1857 bis 1871* (Berlin: Ernst Siegfried Mittler und Sohn, 1900), 49

of the needle-rifle in 1866, would the Austrian shock columns have fared any better than the Russian columns at Inkerman or the sepoy formations in the Indian Rebellion?

When Prussia accepted the *Zündnadelgewehr* in 1841, the remarkable weapon was supposed to be a state secret and initially the rifles were stored in armories, to be issued only in a national emergency. Compared to the weapons of the 1840s, the *Zündnadelgewehr* was truly a force to be reckoned with. It predated the general adoption of the Minié rifle by many years, although the slow rate of production meant that the new needle-rifle was not available in sufficient numbers for the Prussian Army until the revolutions of 1848. Every other European power of the 1840s armed their infantry with a variation of the smoothbore musket, effective only at short range and capable of being fired (at best) three times a minute. The *Zündnadelgewehr*, on the other hand, was a rifle. As a breechloader, the heavy ball it fired was loaded easily at the breech, and forced into the rifling grooves when fired. Out to 200 or 300 yards, the *Zündnadelgewehr* was reasonably accurate and could be fired six times a minute or even faster; it does not take much imagination to consider the enormous advantage this presented when facing enemy forces equipped with inaccurate smoothbore muskets.

By the mid-1850s, however, the wonderful *Zündnadelgewehr* was actually obsolete. All major European powers had adopted rifle-muskets by 1855. These new rifled muzzleloaders had a higher muzzle velocity, flatter trajectory, and much longer effective range than the *Zündnadelgewehr*. In some ways, the tables had been turned on the Prussians. Instead of Prussian troops pouring effective rifle fire into helpless smoothbore-armed enemies at 250 yards, there was a very real possibility that *Zündnadelgewehr*-armed Prussians might someday be taking effective rifle fire at 600 yards from enemy rifle-muskets. In the early 1850s, when the British Small Arms committee was searching for the ideal rifle to arm the infantry of the British Army with, they acquired Prussian-made examples of the *Zündnadelgewehr*, and even made a few English needle-rifles at the government factory at Enfield. Ultimately, the committee decided that rate of fire wasn't everything, and conservatively chose to stick with muzzleloaders. Extensive tests revealed that the unique half-ball half-cone shaped bullet fired by the *Zündnadelgewehr* was woefully inferior to the ballistics of new conical muzzleloading rifle-musket bullets, and the British chose instead to adopt the Pattern 1853 Enfield and its cylindroconical Pritchett bullet. In 1854, the Austrian Empire adopted the new Lorenz rifle, which was cutting-edge technology for its time with a small .54-caliber bore, and excellent ballistics from a light and very accurate bullet. With a muzzle velocity approaching up to 1300 feet per second, the Lorenz had a flatter trajectory and longer dangerous space than any contemporary military rifle. The Lorenz rifle had the same length of dangerous space at 500 yards that the *Zündnadelgewehr* had at 300.

Somewhat desperately, in 1855 the Prussians resorted to a stopgap measure to improve the ballistics of the needle-rifle. The heavy sphere-cone bullet became increasingly inaccurate at ranges beyond 300 yards, and was discontinued. Instead, an acorn-shaped 13.5mm *Langblei* (literally "long lead") bullet was fitted to a pressed paper sabot in the M/55 version of the needle-rifle's paper cartridge. The bullet itself never actually touched the barrel, and the greased paper sabot produced less friction, increasing the muzzle velocity to 970 feet per second. With an elongated shape, the *Langblei* had marginally better aerodynamic qualities over the old bullet, even though the unique acorn shape was not ideal for a projectile and was purely necessary to ensure that the sabot separated from the bullet after leaving the muzzle. About 10% of bullets were not seated concentrically in the sabots in the hand-made paper cartridges, causing them to fly wild. Additionally, a small but appreciable percentage of sabots failed to separate from the bullet, with ruinous effect on accuracy.[220] The Prussian soldiers even had a name for the rotary buzzing sound the unseparated bullet and sabot made when flying wildly

---

[220] Eckardt, Werner and Morawietz, Otto, *Die Handwaffen des brandenburgisch-preussisch-deutschen Heeres 1640-1945* (H.G. Schultz, 1957), 109

downrange: *Brummer*, or bluebottle fly. While the *Langblei* cartridge marginally increased the muzzle velocity and effective range of the *Zündnadelgewehr*, it remained inferior to the latest models of rifle-musket being adopted by the European powers in the late 1850s in every respect, with the sole exception of rate of fire.

Captain W. Vogel, a Prussian officer writing just a year before Königgrätz in 1865, cautioned that the *Zündnadelgewehr* was no longer to be considered superior to the latest generation of Minié rifles in service across Europe. Especially when used in an intelligently conducted defense, the Minié rifle's flat trajectory and longer dangerous space posed a serious risk to Prussia infantry with needle-rifles. "Where only long range and certain shooting is concerned, the Minié rifle has equal or perhaps even greater advantages," Captain Vogel wrote in his book *Das Preußische Zündnadelgewehr*. "But if [soldiers with the Minié rifle] inaccurately estimate the distance, the Minié falls behind the *Zündnadelgewehr*, which compensates for its inferior accuracy by greater mass of fire." The needle-rifle would demonstrate its superiority only at relatively close range on the open field, under equal conditions, where enemy formations with Minié rifles were deprived of their advantages of long range or prepared defensive positions. "In two minutes, the volleys of the *Zündnadelgewehr* compared to the Minié rifle would be 9 to 3, and with rapid fire (*Schnellfeuer*) the ratio would be even more unfavorable to the Minié."[221]

The rate of fire of the *Zündnadelgewehr* could itself be a disadvantage if soldiers expended their ammunition too soon, or by firing ineffectually at long ranges. This was a great concern even with muzzleloading rifles, and skeptical British authorities were constantly worried that jittery soldiers would fire off their ammunition unless kept under strict fire discipline and control by their officers. In 1864, Prussian infantry actually did run out of

---

[221] Vogel, W., *Das Preußische Zündnadelgewehr und seine Vorzüge, sowie die verbesserten Handfeuerwaffen der Infanterie überhaupt, nebst Beiträgen zur Theorie des Schießens* (Eduard Döring, Potsdam: 1865), 84

ammunition during battles in Denmark, but there were sufficient reserves available to avoid a catastrophe. Moltke realized that the *next* war would probably be against a major European power, and an outnumbered Prussia would not have available reserves to plug in for units out of ammunition.[222] The horrible trajectory of the slow-moving *Langblei* made accurate long-range shooting even more difficult than with a rifle-musket like the P1853 Enfield. Moltke wrote in his 1865 *Remarks* that "due to the very difficult estimation of distances to the target, it is thought that in 20 minutes the unsupervised soldier can undoubtedly fire his 60 cartridges, especially in the heat and excitement of battle." The needle-rifle's rate of fire was "a tremendous means of success in the hand of the leader who knows how to bring it to bear at the right moment, [but] it is simultaneously a danger to the uncontrolled individual."[223]

Yet Moltke remained committed to *Feuertaktik* in spite of the tactical problem the needle-rifle created. He studied the decisive effect of rifle fire in the Crimean War, particularly at Inkerman where the Russian shock columns sustained "enormous fire-losses" (*Feuerverluste*).[224] After digesting the lessons of Solferino and Magenta in 1859, Moltke doubled down on *Feuertaktik* and spent the next several years aggressively lobbying the reluctant Liberal Prussian diet to fund comprehensive marksmanship instruction for the entire Prussian Army. In 1862, with the help of the recently-appointed Prime Minister Otto von Bismarck, Moltke rammed through a major army reform that included a substantial increase in the quantity of ammunition allocated for rifle training. Nearly a decade after the British Army had mandated musketry instruction for every soldier (regardless of

---

[222] Wawro, *The Austro-Prussian War*, 23
[223] Moltke, *Remarks*, 52
[224] It is particularly revealing that in his Remarks of the Year 1865 on the Influence of Improved Firearms in Tactics, Moltke did not mention the American Civil War even once; the thunderous silence of this omission speaks volumes.

unit type or designation) with 110 rounds for new recruits and 90 rounds for trained soldiers fired each year, the Prussian Army finally allocated 100 rounds per year for rifle training. Compared to previous allocations, and compared to other continental powers in general, 100 rounds a year was, as Dr. Wawro describes it, a veritable "shooting spree."[225] Moltke led the effort. By 1865, he appreciated the valuable fact that "every single soldier is trained at precision shooting" (*Feinscheißen*). Much like General Hay and Colonel Wilford, Moltke also sought to develop the soldier himself. "The soldier is conditioned for real shooting, because the rifle barrel does not rest in a motionless gun carriage," Moltke wrote. "An almost instinctive estimation of the distance, fast aiming, correct acquisition of a moving, living object, and quick decision made with a sure calmness are personal qualities whose absence that even the most superb weapons cannot make up for."[226] The Prussian Army rewarded individual and unit marksmanship with cash rewards; first-class shooters received two Prussian Thalers, and second and third-class shooters received one. The soldier had the option of receiving the reward in cash, or in the form of a medal with one Thaler worth of gold.[227]

One minor skirmish in the Schleswig War of 1864 particularly attracted Moltke's attention. Near the village of Lundby, outside of Aalborg, Denmark, the 124 Prussian soldiers of Captain von Schlutterbach's company of infantry from the 50th Infantry Regiment were engaged in a reconnaissance. It was the morning of July 3, 1864, exactly a year to the day since the climactic end of the Battle of Gettysburg. The I.R.50 was an ordinary line regiment, armed with the standard line infantry Model 1841 needle-rifle; by mid-1864, Captain von Schlutterbach's company would have benefitted from Moltke's army reforms that instituted comprehensive rifle instruction. The

---

[225] Wawro, *The Austro-Prussian War*, 24
[226] Moltke, *Remarks*, 52
[227] "Die Schiessübungen der preussischen Infanterie," *Österreichische militärische Zeitschrift* Vol. 3, 1865, 29

Danes had discovered them, however, and the commander of the Danish 1st Infantry Regiment, Lt. Colonel H.C.J. Beck, personally led a force of 180 troops from his regiment's 5th Company on an aggressive and rather brilliant maneuver around the Prussians. By dawn, the Danes had positioned themselves on a small hill 650 yards behind the Prussian rear, cutting them off from the rest of the Prussian forces in the area.[228]

Immediately, Lt. Colonel Beck exploited his surprise and launched a miniature Jominian attack with his 180 men, formed into a column 10 soldiers wide and 16 deep, the officers and NCOs remaining outside the column for direction and discipline. They moved quickly, following the established tactics of the day that Napoleon III had summarized in 1859: the new rifles can only harm you at a distance, so cover the ground quickly, get out of the dangerous space, and bring the bayonet to the enemy. Moltke noted that the Danes sprung "resolutely charged ahead for the bayonet attack in column formation." The Prussians rapidly took up positions in a convenient ditch, putting 76 soldiers on the firing line and prudently holding back 48 in the rear as a reserve. As the Danes increased their pace as they came closer, Schlutterbach held his fire. His troops were green, and due to the poor trajectory of the *Langblei*, fire at ranges beyond 300 yards was difficult even for experienced troops. Schlutterbach chose instead to hold his fire until the Danes came within 300 *Schritt*, or about 235 yards. Finally, with Lt. Colonel Beck's column bearing down, the order to fire was given along the Prussian line at 250 *Schritt*, according to Schlutterbach's post-battle report.

The massed, near point-blank volley ripped into the Danes; the column staggered but pressed on, as it was designed to do, by the inertia of its own mass of humanity. Losses in the attack by column were expected and accepted, but Jomini's cold arithmetic had been laid down with the assumption that the firing defender would be able, at most, to deliver two volleys a minute. Napoleon

---

[228] Embree, Michael, *Bismarck's First War: The Campaign of Schleswig and Jutland 1864* (UK, Helion & Co 2006), 325

III and his Italian allies had suffered enormous losses with their direct bayonet assaults in 1859, but they carried the day. But only a few seconds after the first volley, more Prussian *Langblei* were tearing into the Danes. Like the sepoys in India seven years earlier, the Danes were learning the hard way about being on the wrong side of a military paradigm shift. Moltke admired their pure courage: "despite receiving the second blast of fire, they continued their way with enormous bravery" (*mit großer Bravour*).[229] It was the same enormous bravery that had driven the Austrians out of their positions at Magenta and Solferino, but it could not overcome the unprecedented volume of fire that poured from the Prussian position along the ditch. The column fell apart and the Danish cohesion failed. Schlutterbach brought his reserve up, which added their fire to the others. None of the Danish bayonets got anywhere near the Prussian ditch, and the Danes who had survived the destruction of the column broke and retreated. Out of 180 Danish soldiers who had gone forward in the attack, 32 of them were killed outright and another 66 wounded or captured, for a total loss of over 50% of their strength. Three Prussian soldiers were wounded.

There is some debate over how the Prussian fire was delivered. The Danes believed they were controlled volleys; the official Prussian record describes the fire as volleys (*Salven* or "salvoes" in the German). Captain von Schlutterbach said he fired the first volley and then went immediately to *Schnellfeuer* (and he tellingly used that word to describe it). The historian Michael Embree points out that the Prussians were so rigidly drilled that, after the first volley, "they individually loaded almost in time with each other."[230] Moltke wrote a year later that there had been four volleys. With Teutonic efficiency, the meticulous Prussians recorded that 750 cartridges had been expended by the defenders. Considering that a third of the Prussian force had been held back as a reserve, and did not engage until the Danes had gotten precariously close to the ditch, it can be safely determined that

---

[229] Moltke, *Taktisch-Strategische Aufsätze*, 58
[230] Embree, 328

the Prussians on the line were rapidly loading and firing.

Lundby was a minor skirmish, even by Schleswig War standards, but the lesson was recognized and absorbed by Moltke and the Prussian advocates of *Feuertaktik*. A numerically superior force had been nearly annihilated by musketry fire alone. Junior officers had successfully controlled and directed the fire of their soldiers. The rapid fire of the *Zündnadelgewehr* meant that *Feuertaktik* could be successfully employed at relatively close range. In the British Army's more primitive version of fire tactics, great emphasis was placed on training soldiers to deliver effective fire at long ranges out to 600 yards and beyond. It was necessary to fire with effect at long range because an attacking force could cross the dangerous area and bring the fight to the bayonet before soldiers with muzzleloaders could fire very many rounds. The Prussians understood that they didn't *need* to fire at longer ranges, because the *Zündnadelgewehr* could dispense such a great volume of fire in a comparatively short time. Pure *Feuertaktik* could destroy an attacking enemy, even if fire was withheld until the enemy was within 200 yards.

For Moltke, 300 *Schritt,* or about 230 yards, was the golden distance for opening fire. He refers to 300 *Schritt* repeatedly across his prolific writings from 1865 through the Franco-Prussian War of 1870-71. As he wrote in his *Remarks,*

> The lack of success of long-range fire deprives soldiers of the confidence in the immediate effect of close fire, and the most dubious consequences could occur. A deployed battalion, capable of engaging an enemy up to about 300 *Schritt*, will still have time to send thousands of bullets towards the opponent, even against the most determined sort of enemy, before proceeding to the bayonet attack itself. And likewise, a commander who knows how to bring his battalion to 300 *Schritt* of the enemy, can count on sure success. The moment will not be long in coming where he can launch a real bayonet attack if the opponent, after very

large losses, is even willing to resist.[231]

The objective of Moltke's strategy was to simultaneously concentrate Prussian armies around a "pocket" (*Kessel*)of enemy forces; similarly, the tactical objective was to envelop an enemy formation and smother it with fire at medium to close range. Moltke realized, after the lessons of 1864, that Prussian infantry could destroy any enemy force with pure *Feuertaktik* as long as they could get within 300 *Schritt*. Then, beneath a hail of *Schnellfeuer*, the enemy would be shot to pieces, disorganized, and then the position could be easily taken. When every other European army (with the lone exception of the British Army) was wholly committed to shock tactics and spirited bayonet assaults by courageous, confident soldiers driven by *elan*, Moltke saw the paradigm shifting before his eyes. "Even the most brilliant bravery fails before resistance that cannot be surmounted," Moltke cautioned Prussian infantry leaders in 1865. "The bayonet attack should not be seen as the first recourse in battle, but rather the last."[232] More to the point, "the combat powder of infantry rests on the effect of its fire. Its success depends on attaining fire superiority and exploiting it decisively and rapidly." Just as General Hay and Colonel Wilford had realized a decade before, Moltke emphasized that "individual marksmanship training is of the utmost importance."[233]

As part of the Prussian Army reforms, Moltke broke up the large infantry regiment as the main tactical element and, plowing through objections from old relics like Wrangel, introduced new small unit tactics designed around *Feuertaktik* and the needle-rifle. "When von Moltke analyzed the battles of Alma and Inkerman," Dr. Martin Samuels explains in study of British and German tactics leading up to the First World War, "he concluded that the

---

[231] Moltke, *Taktisch-Strategische Aufsätze*, 54
[232] Ibid., 59
[233] Hughes, Daniel J., tr., *Moltke on the Art of War; Selected Writings* (Presidio Press, Novato, California: 1993), 154

Russians had over-emphasized shock tactics and that a certain dispersal was necessary in the face of modern fire," from Minie rifles.[234] The Prussian battalion column was split up, as Dr. Wawro describes, into "nimble rifle companies and platoons" that operated with a great degree of independence, at least by 1860s standards. The *Zündnadelgewehr* provided the greatest advantage when small units operated in skirmish formations, bringing every possible rifle to bear. Junior officers, down to the platoon level, were trained in fire control and taught when to recognize the decisive moment of an engagement, and only then order their soldiers to commence an independent *Schnellfeuer*. Only with effective command and control could soldiers be prevented from firing off their precious cartridges rapidly at long ranges. "When it comes to negotiating the deciding moment," Moltke wrote, "any consumption of ammunition is justifiable, but it must always remain the responsibility of the leaders to designate these moments."[235]

These tactical necessities led to Moltke's greatest contribution to the modern operational art: *Auftragstaktik*. Usually translated as "mission command," it can be broadly described as a commander providing subordinates with his intent and mission objectives, and then allowing the subordinates to execute the order by achieving the mission objectives in the best way possible. Previously, the commander (like a Napoleon or a Wellington) was personally able to observe the battlefield, and direct the battle order by order. As armies grew larger in the industrial age, Moltke realized it would not be possible for a commander to view the whole battlefield. Nor was it practical for a large unit commander to exercise tight command and control of his troops, if they were broken down into small units to exploit the rapid-fire capability of the *Zündnadelgewehr*. Napoleon won his battles by personally directing where and when to launch the

---

[234] Samuels, Martin, *Command or Control?: Command, Training and Tactics in the British and German Armies, 1888-1918* (Frank Cass, London: 1995), 64
[235] Hughes, 196

sledgehammer-like blows of his shock columns, "under the watchful eyes of the emperor himself."[236] Moltke would win his battles by issuing mission objectives, allowing the junior officers who had "boots on the ground" the flexibility to improvise and adapt to the reality of the battlefield, as long as they worked toward the mission. Dozens or even hundreds of company and platoon-sized elements, working with a "common operating picture," would maneuver and outflank the clumsy enemy columns, then smother the enemy formations with *Schnellfeuer*. While brilliantly simple and obvious to us today, it was outrageous and heretical to the military establishment of the time. Most authorities outside of Prussia expected Moltke's crazy system to fail spectacularly, with all order, organization, and discipline lost as lieutenants wandered around in confusion with their platoons. Instead, it was a spectacular success and *Auftragstaktik* would remain the centerpiece of Prussian, and later German, military doctrine until 1945.

Under Moltke's direction, the Prussian General Staff analyzed and digested the experience of 1864 and made mobilization plans for the next war. Meticulously planned railroad operations would haul vast numbers of reservists across the Kingdom of Prussia to the rallying points, in record time. Everything depended upon outmaneuvering a slower, clumsier enemy army, and annihilating it in a quick war-ending decisive battle. Moltke and the Prussians (and later the Germans in both world wars) would pursue this fleeting goal of early decisive battle because of the legacy the small Prussia, a "kingdom of border strips," left on the German military profession. Prussia's limited manpower and industry would not support a long war of attrition. Strategic speed, mobility, envelopment, and concentration would bring the Prussian armies to the battlefield; tactical innovation, *Feuertaktik*, *Schnellfeuer*, and *Auftragstaktik* would overwhelm the enemy in overlapping fields of fire delivered by swarms of small units, all working toward a mission objective with soldiers highly trained (by 1860s standards) at

---

[236] Wawro, *The Austro-Prussian War*, 19

individual marksmanship.

Meanwhile, the army of Emperor Franz Josef in 1866 was among the largest and most experienced professional armies in Europe. The Austrian Empire had the recent lessons of the major campaigns and field battles of 1859 to draw from, as well as the 1864 campaign alongside Prussia in the Schleswig War. In sharp contrast to Moltke and the Prussian Army, the Austrians completely rejected *Feuertaktik* and committed almost wholly to *Stosstaktik* after the disaster of 1859 at Magenta and Solferino. According to new regulations issued in 1862, the primary function of infantry was to "deliver the mass assault with the bayonet."[237] Arguably the best rifle-musket in Europe, the M1854 Lorenz with its high muzzle velocity, long range, and highly accurate bullet, was relegated to a convenient pole upon which to mount a bayonet. Instead of greater dispersion of soldiers on the battlefield, the Austrians shoved their assault column battalions even closer together. Unfortunately for thousands of Austrian soldiers, who would die in heaps as their columns charged into withering Prussian *Schnellfeuer* at Königgrätz, the Austrian employment of *Stosstaktik* in 1864 against the Danes only served to confirm the superiority of shock tactics. The parsimonious Habsburgs slashed the annual allocation of cartridges for shooting practice to a mere 20 rounds, while the British were firing 90 per year and the Prussians 100.

The most comprehensive scholarly attention given to Austria's embrace of *Stosstaktik* is Geoffrey Wawro's journal article *An Army of Pigs*. In the immediate aftermath of 1859, the Minister of War, General August Degenfeld-Schonburg, obtained the willing approval of the Emperor to condense the dispersed Austrian formations of the 1850s into a dense mass. The space between half-battalion formations was shrunk from 54 to 12 paces, and they were frequently joined together into a mass battalion column. Dr. Wawro described the formation as a "death-trap." The Emperor and Degenfeld "embraced shock and

---

[237] Rothenberg, Gunther, *The Army of Francis Joseph* (Purdue University Press: 1988), 64

phased out fire tactics [because] they sincerely believed that small arms fire had been counterproductive in 1859." After 1864, "the Austrian high command – the Emperor, his adjutants, Benedek, the General Staff, and the war ministry – insisted that Austria had won in Denmark for the very same reasons the French had won in 1859: dash and pluck, nothing more."[238]

In 1863, an Austrian officer identified only as "Captain L. K." wrote a short article that appeared in the Austrian military's professional journal, *Österreichische militärische Zeitschrift,* defending shock column tactics.[239] In the piece titled "The Line Fire of Infantry against Columns," Captain K. declared that "the fire from a deployed line against a column attacking it will, in most cases, be far less effective than hoped for." Almost laughably, Captain K. explains *why* the fire from the line would be ineffective: "In line fire delivered directly to the front, most often the shots will be aimed at the head of the column, which is directly in front as the column approaches, while the other shots pass to the left and right of the column." While the front of the column is exposed to fire and will take casualties (an accepted fact among the school of *Stosstaktik*), the rest of the column will "have little to fear," because oblique fire into the flanks of a column "is seldom found in reality." Even with the clarity of historical hindsight, it is still painful to read these words so confidently written by a junior officer with complete faith in the dangerously and negligently obsolete tactics and doctrine of his service. Captain K. imagined a column attacking a defensive line that would myopically fire only at the very front of the column, and that most of the bullets would harmlessly fly to the left or fight. With simple confidence, Captain K. dismisses the fire of trained soldiers with rifles. "As for marksmanship, you have to doubt it when the movement of the column and overshooting

---

[238] Wawro, *Army of Pigs,* 429

[239] It was very common, in the periodicals of the 19th century, for contributors to make submissions anonymously. Often, the context or particulars of the submission revealed the identity of the author.

[*häufige Überschiessen*] are considered."[240]

There were other non-tactical motivations for the wholehearted Austrian adoption of shock tactics. The Austrian Empire held dominion over dozens of ethnic groups, and after mobilization an Austrian field commander would find battalions of Croats, Slovenes, Serbs, Hungarians, Romanians, Poles, Ukrainians, and Czechs in his army. Few of the peasant conscripts understood more than a few rudimentary words of German, at best. These multi-ethnic units had to be kept together in close formations, lest the battlefield turn into a hopelessly confused mix of soldiers unable to understand each other or comprehend the orders of officers. Unlike the British Army's long-term professional soldiers, the Austrian Army conscripts were quickly trained during a relatively short period of active duty, and then to save money, were trundled off to the reserves to form a large manpower pool of soldiers in event of mobilization. *Feuertaktik* and musketry training took time, money, and professional instruction; *Stosstaktik* and massed columns of soldiers were cheap, easily controlled and directed, and (from the experiences of 1859 and 1864) apparently very effective. Dr. Wawro argues that "the motley social fabric of the Austrian regiment induced the Austrian command to adopt the crudest, simplest tactics in order to conserve ammunition and avert a breakdown in communication."[241] Ludwig von Benedek, the hero of Solferino who would command the Austrians at Königgrätz, distilled the Austrian doctrine in his general order to his army in 1866: "March with resolute attitude to within 300 *Schritt* of the enemy. Then advance at the double and rout him."[242]

Clever diplomatic maneuvering by Otto von Bismarck compelled the Austrian Empire to declare war on the Kingdom of Prussia over the administration of the Schleswig territories

---

[240] "Das Infanterie Linienfeuer gegen Colonnen," *Österreichische militärische Zeitschrift* Vol. 3, (1863), 383
[241] Ibid., 408
[242] Ibid., 430

seized from Denmark in 1864. In the brief seven week war, the Prussians routed the Austrians in a major field battle. The Prussian victory has been credited as the (obvious) triumph of the modern rapid-firing breechloading rifle over the obsolete and slow muzzleloader. While the needle-rifle *helped*, Königgrätz was instead the triumph of modern *Feuertaktik* over obsolete *Stosstaktik*. To make the counterfactual argument, the Prussians could have been armed with the British P1853 Enfield instead of the *Zündnadelgewehr*, and the battle would have largely unfolded the same way. Prussian casualties may have been slightly higher, and Austrian casualties slightly lower, but at the end of the day the Austrian shock columns would have been riddled with Prussian fire. In some cases, a rifle-musket may have been preferable instead of the *Zündnadelgewehr* with its poor trajectory. Austrian massed columns, even at 600 yards, are hard targets to miss; a soldier could be woefully incorrect in his estimation of the range, and still drop a bullet in the rear of the column. The simple and undisputed fact about Königgrätz is that the Austrians used the *worst* conceivable tactics against an enemy utilizing *Feuertaktik* with modern rifles. The Austrians crawled (or, rather, charged enthusiastically) into their own coffin; the *Zündnadelgewehr* was just one of many nails that hammered the lid shut.

The first engagement between Prussian and Austrian troops in the war compellingly demonstrates that it was Prussian tactics, more so than the amazing rate of fire of the needle-rifle, that won the day. Perhaps equally important, the Austrian tactics contributed to their own defeat. On June 26, 1866, there was a one-sided skirmish at the village of Hühnerwasser. A battalion of mixed Hungarian and Romanian line infantry surprised a few Prussian companies outside the town. After an initial inconclusive exchange of potshots, the Austrians launched a bayonet charge with a pair of assault columns. Four Prussian companies, formed in loose firing order, kept their remarkable fire discipline and their officers waited for the Austrian columns to cross the invisible line at 300 *Schritt*. Once they did, the order was given and the Prussian volley stopped the Austrians almost in their tracks. Entire ranks of the Austrian columns had crumpled into bloody heaps; as the Prussians reloaded, Austrian officers could be seen in the distance trying to gather up the disorganized columns and drive them on towards the Prussians. A second Prussian volley broke the Austrian columns, which disintegrated into shaken, fleeing soldiers who turned to the rear and ran. No Habsburg bayonets came within a hundred yards of the Prussian line, and the path of the assault columns was strewn with dead and wounded. While the Prussians had lost about 50 men from Lorenz rifle fire in the initial skirmish, the Austrians lost 13 officers and 264 men. This was the work of only two aimed volleys from a few Prussian companies at 300 *Schritt*. The order had never been given to commence *Schnellfeuer,* as the Austrians had never gotten close enough.[243]

The needle-rifle would not have to wait long to demonstrate its rapid-fire capability. That evening Prussian and Austrian troops fought a savage close-range battle through the streets of Podol, fighting house to house and from street barricades. As night fell, the Prussians followed their doctrine and sent 400 infantry on a wide movement in an attempt to envelop the Austrians in the Podol pocket. Instead, the Prussians ran into *thousands* of troops of two full Austrian battalions, that had been held in reserve. The Austrians attacked at once in shock columns. The Prussians, in line formation, fired relentlessly into the Austrian masses. This time, the order for *Schnellfeuer* was passed down the line and the Prussians loaded and fired independently. Three times, the Austrians charged in column towards the "thin blue line" of 400 Prussian infantry, only to be repulsed with heavy loss. After expending all their ammunition from sizzling hot rifles, the Prussians abandoned their attempt at envelopment. They had inflicted over 1000 casualties, and repulsed determined assaults by enemy forces five times their number. *Feuertaktik* had defeated *Stosstaktik* twice on the same day. For the courageous yet doomed Austrian infantry, it would not get any better.

For the next week, the Prussian and Austrian armies fought

---

[243] Wawro, *The Austro-Prussian War*, 129-130

a series of minor, sharp engagements, of which the Austrians only "won" one, at Trautenau on June 27. While the Habsburgs found themselves in command of the field, it had cost them nearly 5,000 casualties to 1,500 Prussians. In virtually every instance, *Stosstaktik* and the shock column had failed, with Austrian battalions falling to pieces before Prussian *Feuertaktik*. And yet in the face of repeated failure, the assault columns went forward and the Prussians shot them down. Benedek himself, only days after his general order that directed commanders to get within 300 *Schritt* of the enemy and then charge at the double, was shocked by the enormous casualties. On June 28, Benedek dispatched a new order attempting to refocus Austrian tactics around the excellent rifled Austrian artillery, which had proven highly effective. "The not insignificant losses suffered by the infantry have convinced me," Benedek ordered, "that the infantry should go into action, and especially into the bayonet charge, only after the enemy has been shaken by artillery fire."[244] It was too little, too late. On the 29th, an Austrian corps was smashed at Jicin with crippling losses, and Benedek fell back to Königgrätz on the Elbe River. He famously wrote to the Emperor to beg him to seek peace terms with Prussia, and described the campaign as a "catastrophe" that could only result in defeat.

A few days later, at Königgrätz, three Prussian armies enveloped the Austrians. Moltke's concentration was brilliant, but not perfect, and he was gravely worried that the Austrians might overwhelm one of the Prussian armies before the others could arrive. At several points of the battle, Prussian victory was far from certain. King Wilhelm I of Prussia had arrived on the battlefield, alongside Moltke and Bismarck, and was watching a crisis unfold. The Austrians, true to their doctrine, were launching aggressive assaults at desperately outnumbered Prussian elements, who were hanging on precariously and waiting for the rest of the Prussian armies to concentrate. The situation was so bad that, before noon on July 3, King Wilhelm I turned to his and murmured, "Moltke, Moltke, *wir verlieren die Schlacht!*"

---

[244] Rothenberg, 70

(we are losing the battle) and then asked about what preparations were being made for an orderly retreat. Moltke, however, peering through a telescope, saw the dusty vanguard of the Prussian Second Army approaching in the distance, accomplishing his concentration and envelopment. The king was convinced that their position was secure, and the battle ended, ultimately, in the famous Prussian victory.

It is in the combination of fire tactics and decentralized, loose order formations that set the Prussians apart from, and ahead of, the British. While the British Army had adopted a more primitive form of fire tactics, it was without the prototypical *Auftragstaktik* and loose formations of the Prussians. In the British infantry regulations of 1859, the old close order maneuvers in lines of tightly-formed infantry were retained. Fire was to be delivered primarily from the line, just as Wellington had done at Waterloo. By the early 1860s, the British were straddling both sides of the shifting paradigm. All British infantry soldiers were expected to be able to perform as light infantry, regardless of the designation of their regiment. Every British soldier was armed with the rifle, fitted with precise long-range sights, and every soldier received (in theory) the same extensive musketry instruction, and was subjected to the same annual practice and qualification. Year after year, every soldier expended between 110 and 90 rounds in practice. The British Army, in the 1859 regulation, expected to fight in the old conventional formations, but anticipated the formations opening fire upon the enemy from long distance, and defeating the enemy with fire before he could close in.

Against an enemy using shock tactics, this was seen as a prudent and reasonable mode of combat. Keeping the soldiers in ranks, in close order, maximized the available firepower of the muzzleloading Pattern 1853 Enfield into the smallest space. Soldiers would be encouraged by the immediate presence of their comrades in a strong line. Open, extended light infantry order could not generate the firepower necessary to stop the powerful shock columns or the agile *Chasseurs* that moved rapidly across the battlefield. Tight formations delivered crushing volumes of fire during the Crimean War and Indian Rebellion. The role of light infantry retained its role as a screening force of skirmishers, but the line remained preeminent in the decisive culmination of battle. Colonel Wilford and General Hay, for all their enthusiasm for the rifle, for fire tactics, and marksmanship, imagined British troops standing in line on the battlefield, pouring long-range aimed fire into enemy columns. They always assumed Her Majesty's enemies would still employ smoothbore-era tactics, like the Russians and Sepoys, and they never developed tactical theories for fighting enemies who were themselves armed and proficient with the rifle.

The rifle-musket started the paradigm shift; it could not complete it. While it was the first modern infantry weapon, it was very quickly superseded by breechloading rifles that would build upon the foundation the rifle-musket had laid. It *did* revolutionize the battlefield, provided (of course) that the soldiers using it had received *the* paramount mark of modern soldiers: adequate individual training in their arms and tactics. The rifle-musket demonstrated that firepower was superior to shock, that the individual trained soldier was a valuable asset, and that fire tactics were viable alternatives to Napoleonic and Jominian masses, jogging forward with bayonets. Yet it did this simply by virtue of increasing the range and accuracy that the average infantryman could fire his approximately three rounds per minute, a rate of fire unchanged since the first introduction of the flintlock smoothbore musket almost 200 years earlier. The paradigm was shifted, and trained soldiers with the rifle-musket were able to inflict defeats upon enemy forces who fought under the construct of the old battlefield ways, time and time again.

Even so, the rifle-musket was still a *musket*. For all the changes it inaugurated in how soldiers were trained, and how British troops would use firepower to fight future enemies on the battlefield, the soldier still had to shove the bullet down the barrel of his weapon with a long metal stick. Colonel Wilford himself might have formed an entire picked battalion of Hythe-trained first-class shots with the P1853 Enfield rifle, and a lone Prussian rifle company with Dreyse needle-rifles would have been able to put the same number of hits on a target 300 yards away in an equal span of time. The Prussians also demonstrated that the chief obstacle to the adoption of "rapid fire" arms – that soldiers might fire away their ammunition recklessly and leave them vulnerable – was overcome with proper training, empowerment of junior leaders, and *Auftragstaktik*. The rifle-musket was hindered by a low rate of fire and, correspondingly, a relatively low volume of proportional firepower produced by an infantry unit. Because of the slow rate of fire, dense close order lines of infantry were reluctantly retained in order to provide the fire control and the raw numbers of rifle-muskets necessary to generate sufficient firepower to defeat an enemy using fire tactics. In this role, it was woefully inferior to the breechloading firearms that were proving to be as reliable, effective, and accurate as any muzzleloader by the start of the 1860s.

13
CONCLUSION

The United States may have come dangerously close to war with the United Kingdom in early 1862 in the aftermath of the Trent Affair. A pair of Confederate envoys, en route to Paris and France, were forcibly removed by a U.S. warship from the Royal Mail Ship *Trent* in December of 1891. Exactly *how* close things came to actual hostilities is quibbled over by historians, but the potential for real conflict was there. While the British demanded an apology and a release of the Confederate envoys, U.S. Secretary of State William Seward refused to back down. William Howard Russell, the famous *Times* correspondent who described the rifle-musket as "the Destroying Angel" and documented the Charge of the Light Brigade, ran into Seward in Washington at a diplomatic function, during the height of the Trent Affair. Russell supposedly overheard a somewhat inebriated Seward boast that, if Britain really wanted war, "We will wrap the whole world in flames! No power is so remote that she will not feel the fire of our battle and not be burned by our conflagration!"[245] As relations soured between the countries, the British began quietly reinforcing the Dominion of Canada. Among the thousands of British troops sent to Canada was, ominously, Colonel Lane-Fox, the writer of the original *Instruction of Musketry* regulation and the expert musketry instructor who trained the British regiments at Malta on their way to the Crimean War.[246] Once again, he was tasked with training British troops that were possibly on their way to war; the troops of the New York State Militia at Fort Niagara could hear the distant musketry of the British infantry at practice

---

[245] Burlingame, Michael, *Abraham Lincoln: A Life* (John Hopkins University Press: Baltimore, MD, 2008), 224
[246] Bowden, Mark, Pitt Rivers: The Life and Archaeological Work of Lieutenant-General Augustus Henry Lane Fox Pitt Rivers (Cambridge University Press: 1991), 20

on the other side of the river. Today their expended bullets are dug up on the former shooting-ranges by relic hunters with metal detectors and inexplicably sold on eBay as "Civil War" bullets. Fortunately, cooler heads prevailed and Abraham Lincoln famously advanced his prudent policy of "one war at a time." It is entirely within the realm of the counterfactual historian to speculate on the "what if's" of an Anglo-American War of 1862. That said, if a British army of trained professionals had been able to engage an American army of New York volunteers on a battlefield probably near Buffalo, pitting British fire tactics against the *Chasseur*-inspired American shock tactics, I cannot imagine a favorable outcome for the Americans. The British could have opened fire at 900 yards at the "area targets" of US infantry columns, and maintained their fire indefinitely with high quality Enfield ammunition. Like the Russians in Crimea and the Sepoys in India, the American formations would be receiving effective fire at unfathomable ranges, and unable to give any effective reply. Such a battle would have put to rest, for all time, the debate over the rifle-musket and its transformative influence on the modern soldier. But it was not to be. And so the historians continue to argue about the impact of the rifle-musket.

In the historiography of the American Civil War, we have essentially executed a U-turn in the last 75 years. The common narrative of mid-20th century historians was that the rifle-musket with its long range "Minnie Ball" was responsible for the enormous, unprecedented carnage of that war. Paddy Griffith inaugurated the swing of the pendulum in the other direction, observing the inability of poorly trained soldiers in the American Civil War to use modern weapons to the full extent of their capabilities. The American Civil War was many things: very large, very long for wars of its period, enormously destructive on a scale unseen in the Western Hemisphere. What the American Civil War was *not* is an ideal context for evaluating the effectiveness and impact of the rifle-musket. Instead, the Civil War must be studied in its own unique context. Trained soldiers, even with minimal instruction, successfully used the rifle-musket's new capabilities in the Crimea, India, and elsewhere; untrained volunteers, no matter how enthusiastic, fell back upon using the rifle-musket like the old forgiving smoothbore, both in the Italian wars of unification and the American Civil War. Contemporary historians like Dr. Guelzo and Dr. Hess have denigrated the rifle-musket so completely that the smoothbore musket appears the all-around better weapon. A full analysis of the rifle-musket, in all the contexts in which it was used during the exceptionally brief period from about 1850 to 1865, reaches a consistent conclusion: the rifle-musket's effectiveness depended primarily on the extent and quality of training that the soldier using it received.

The rifle-musket was an imperfect transitional weapon. It required considerable training to employ it with accuracy beyond 200 yards. With a slow rate of fire that was no faster than the older smoothbore musket, and a slow moving heavy bullet that dropped quickly, only trained and instructed soldiers could expect to use it effectively on the battlefield. The rifle-musket *itself* did not spontaneously usher in an era of revolutionary dramatic change on the battlefield. What made the rifle-musket the first modern infantry weapon is, ironically, the inherent difficulty to use one effectively. It brought capabilities that required carefully developed skills to utilize. It required soldiers to be trained, and elevated the soldier from the Napoleonic automaton stuffed in a column with a bayoneted musket into an intelligent, thinking, skilled rifleman. Modern soldiers effectively use weapons of precision, like the rifle, only after being trained: the rifle-musket was the first such general-issue weapon of precision to fit this description of "modern weapon." With the rifle musket, ordinary private soldiers had to become professionals. An untrained volunteer or uneducated conscript could be taught to load and fire Brown Bess in a few minutes, but to get any advantage out of a rifle the soldier needed much more instruction. The *Feuertaktik* of the rifle era meant that soldiers had to be trusted to operate with a degree of independence and autonomy on the battlefield. The rifle-musket, with its great parabolic trajectory, was a clumsy and inefficient bridge between the smoothbore era and the metallic cartridge-firing breechloader, but it forced the Western militaries to adopt new standards of training that today we recognize as modern. The undeniable changes brought about by the rifle in tactics, training, and on the battlefield, certainly qualifies the rifle-musket as the first modern infantry weapon.

ABOUT THE AUTHOR

Brett Gibbons is an Ordnance officer in the U.S. Army and a veteran of Operation Enduring Freedom. He is also an enthusiastic collector and shooter of 19th century firearms, particularly the rifle-musket.

Printed in Great Britain
by Amazon